Dr Luke Heaton graduated with first class honours in Mathematics at the University of Edinburgh before going on to take an MSc in Mathematics and the Logical Foundations of Computer Science at the University of Oxford. After spending a year making mathematically inspired art, he gained a BA in Architecture at the University of Westminster, and was briefly employed as an architectural assistant. He then returned to Oxford, completing a DPhil in Mathematical Biology. He is currently employed by the University of Oxford as a post-graduate research assistant in the Department of Plant Sciences. Heaton's research interests lie in mathematics and the mathematical modelling of biological phenomena, the history and philosophy of mathematics, morphogenesis and biological pattern formation, network theory, biophysics, and the statistical properties of efficient transport networks. He has published several papers on the biophysics of growth and transport in fungal networks.

Recent titles in the series

A Brief Guide to James Bond
Nigel Cawthorne

A Brief Guide to Secret Religions
David V. Barrett

A Brief Guide to Jane Austen
Charles Jennings

A Brief Guide to Jeeves and Wooster
Nigel Cawthorne

A Brief History of Angels and Demons
Sarah Bartlett

A Brief History of Bad Medicine
Ian Schott and Robert Youngston

A Brief History of France
Cecil Jenkins

A Brief History of Ireland
Richard Killeen

A Brief History of Sherlock Holmes
Nigel Cawthorne

A Brief History of King Arthur
Mike Ashley

A Brief History of the Universe
J. P. McEvoy

A Brief History of Roman Britain
Joan P. Alcock

A Brief History of the Private Life of Elizabeth II
Michael Paterson

A BRIEF HISTORY OF

Mathematical Thought

Luke Heaton

ROBINSON

First published in Great Britain in 2015 by Robinson

3 5 7 9 10 8 6 4 2

A CIP catalogue record for this book is available from the British Library.

ISBN 978-1-4721-1711-3

Typeset in Stempel Garamond by Palimpsest Book Production Ltd, Falkirk, Stirlingshire

Printed and bound by CPI Group (UK) Ltd, Croydon CR0 4YY

Papers used by Robinson are from well-managed forests and other responsible sources

MIX
Paper from responsible sources
FSC® C104740

Robinson
is an imprint of
Little, Brown Book Group
Carmelite House
50 Victoria Embankment
London EC4Y 0DZ

An Hachette UK Company
www.hachette.co.uk

www.littlebrown.co.uk

CONTENTS

INTRODUCTION

'Mathematics is the gate and key of the sciences. ... Neglect of mathematics works injury to all knowledge, since he who is ignorant of it cannot know the other sciences or the things of this world. And what is worse, men who are thus ignorant are unable to perceive their own ignorance and so do not seek a remedy.'

Roger Bacon, 1214–1292

The language of mathematics has changed the way we think about the world. Most of our science and technology would have been literally unthinkable without mathematics, and it is also the case that countless artists, architects, musicians, poets and philosophers have insisted that their grasp of the subject was vital to their work. Clearly mathematics is important, and in this book I hope to convey both the poetry of mathematics and the profound cultural influence of various forms of mathematical practice. For better or worse, you can't comprehend the influence of maths until you have some understanding of what mathematicians actually do. By way of contrast, you don't need to be an engineer to appreciate the impact of technological change, but it is hard to comprehend the power and influence of mathematical thought without an understanding of the subject on its own terms.

Most people are numerate, and have learned a handful of rules for calculation. Unfortunately the arguments and lines of reasoning behind these techniques are much less widely known, and far too many people mistakenly believe they cannot hope to understand or enjoy the poetry of maths. This book is not a training manual in mathematical techniques: it is an informal and poetic guide to a range of mathematical thoughts. I disregard some technicalities along the way, as my primary aim is to show how the language of maths has arisen over time, as we attempt to comprehend the patterns of our world. My hope is that by writing about the development of mathematical ideas I can inspire some of my readers, shake up some lazy assumptions about pure and applied mathematics, and show that an understanding of maths can help us to arrive at a richer understanding of facts in general.

Mathematics is often praised (or ignored) on the grounds that it is far removed from the lives of ordinary people, but that assessment of the subject is utterly mistaken. As G. H. Hardy observed in *A Mathematician's Apology*:

> Most people have some appreciation of mathematics, just as most people can enjoy a pleasant tune; and there are probably more people really interested in mathematics than in music. Appearances suggest the contrary, but there are easy explanations. Music can be used to stimulate mass emotion, while mathematics cannot; and musical incapacity is recognized (no doubt rightly) as mildly discreditable, whereas most people are so frightened of the name of mathematics that they are ready, quite unaffectedly, to exaggerate their own mathematical stupidity.

The considerable popularity of sudoku is a case in point. These puzzles require nothing but the application of mathematical logic, and yet to avoid scaring people off, they often carry the disclaimer 'no mathematical knowledge required'! The mathematics that we know shapes the way we see the world, not least because mathematics serves as 'the handmaiden of the sciences'. For example, an economist, an engineer or a biologist might measure something several times, and then summarize their measurements by finding the mean or average value. Because we have developed the symbolic techniques for calculating mean values, we can formulate the useful but highly abstract concept of 'the mean value'. We can only do this because we have a mathematical system of symbols. Without those symbols we could not record our data, let alone define the mean.

Mathematicians are interested in concepts and patterns, not just computation. Nevertheless, it should be clear to everyone that computational techniques have been of vital importance for many millennia. For example, most forms of trade are literally inconceivable without the concept of number, and without mathematics you could not organize an empire, or develop modern science. More generally, mathematical ideas are not just practically important: the conceptual tools that we have at our disposal shape the way we approach the world. As the psychologist Abraham Maslow famously remarked, 'If the only tool you have is a hammer, you tend to treat everything as if it were a nail.' Although our ability to count, calculate and measure things in the world is practically and psychologically critical, it is important to emphasize that mathematicians do not spend their time making calculations. The real challenge of mathematics is to construct an *argument*.

Pythagoras' famous Theorem provides an excellent example of how the nature of mathematical thought is widely misunderstood. Most educated people know that given any right-angled triangle, we can use the formula $a^2 + b^2 = c^2$ to find all three lengths, even if we have only been told two of them. As they have been asked to repeatedly perform this kind of calculation, people mistakenly conclude that mathematics is all about applying a given set of rules. Unfortunately, far too few people can give a convincing explanation as to *why* Pythagoras' Theorem must be true, despite the fact that there are literally hundreds of different proofs. One of the simplest arguments for showing that it's true hinges around the following diagram:

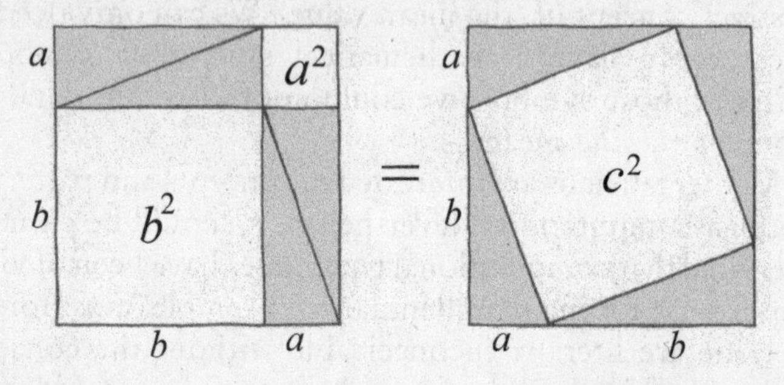

Pythagoras: The shapes on either side of the equals sign are contained inside a pair of identical squares, whose sides are $a + b$ units wide. The one on the left contains a square a units wide, a square b units wide plus four right-angled triangles. The one the right contains a square c units wide plus four right-angled triangles. We can convert the picture on the left into the one on the right simply by moving the four triangles, and moving a shape does not change its area. Since the white area is the same in each of the drawings, this demonstrates that $a^2 + b^2 = c^2$ for any right-angled triangle.

Sceptic: How can you be certain that we always get a square on the right-hand side? More specifically, how do you know that your triangles always meet at a point, whatever the values of a and b?

Pythagoras: Both drawings are of equal height ($a + b$ units high). This tells us that the two triangles that are just to the right of the equals sign must touch at a point, because they only just manage to fit inside the containing square. Similarly, the two triangles on the bottom of the right-hand side touch at a point, because the total length along this side is $a + b$, which equals $b + a$ (the width of the containing squares).

Sceptic: OK, but how do you know that the triangles on the right-hand side always meet at right angles? In other words, how do you know that the shape on the right-hand side is really a square?

Pythagoras: You agree that we have four sides of equal length, and all four corners are the same?

Sceptic: Yes. Rotating the picture on the right by 90°, 180° or 270° leaves the diagram unchanged.

Pythagoras: And despite these facts you still aren't convinced that it's a square? No wonder they call you a sceptic!

My aim in writing this book is to show how the language of mathematics has evolved, and to indicate how mathematical arguments relate to the broader human adventure. This book is related to the work of various philosophers (particularly Ludwig Wittgenstein), but it is not a history of non-mathematical ideas, or an attempt to draw battle lines between conflicting 'big pictures' from the philosophy of maths. I will have succeeded if my writing provokes thought, but I have also tried to argue against the idea that mathematicians discover facts about abstract objects, just like scientists discover facts about physical objects. Mathematical language does not make sense because abstract objects existed before mathematicians! Contrariwise, we can only become cognisant of abstract objects because mathematical language is something that humans can actually use.

It is fundamental to human understanding that our theories or accounts of the physical world are expressed through language. People make *statements* of fact, and the reflective, systematic study of our ability to make statements leads us into the world of maths. Indeed, our understanding of mathematics always begins with a clear, comprehensible case, from which we form a notion of the abstract principles at play. For example, children learn the counting song, and they are then initiated into the practice of counting actual, physical objects. This concrete experience grounds our sense of number, as we abstract away from a particular experience of counting things, justifiably believing that we could set about counting any collection of objects. That is to say, number words become meaningful for an individual as they use those words on some particular occasion, in the presence of actual, countable objects, but once that person has acquired a language, the language itself enables them to think in terms of number, whatever they might wish to count.

Some people mistakenly believe that to do mathematics, we simply need to follow certain rules. I suspect that people arrive at this erroneous position because in order to satisfy their teachers and examiners, all they need to do is apply some rules correctly. In fact, higher mathematics is an essentially creative pursuit that requires imagination. That said, rules are never far behind our creative insights, because in order to contribute to the body of mathematical knowledge, mathematicians need to be able to communicate their ideas. The formal discipline that we require to fully state our arguments is an essential constraint on the shape of mathematical knowledge, but the mathematics that we know also reflects the problems, challenges and cultural concerns that have motivated the various members of the mathematical community.

I hope that reading this book persuades you that mathematicians are explorers of patterns, and formal, logical proofs that can be methodically checked are the ultimate test of mathematical validity. The clarity of a strictly formal proof is a beautiful thing to comprehend, and I think it is fair to say that an argument is only mathematical if it is apparent that it can be formalized. However, while we can gain a sense of understanding by learning to use a particular formal scheme, it is certainly possible to check each step in a formal argument without understanding the subject at hand. Indeed, a computer could do it, even though a computer is no more a mathematician than a photocopier is an artist.

I have taken a more intuitive approach as my aim is not to train the reader in the appropriate formal techniques, but simply to make the heart of each argument as comprehensible as possible. That said, the subject matter of this book is subtle and sophisticated, so there is no escaping the need to take certain arguments carefully and slowly. Mathematics is a subject where you must read the same sentence several times over, and as with poetry, you must read at an appropriate pace.

Over the course of my book I trace out a history of mathematical practice, with a focus on conceptual innovations. I do not claim to have covered all of the key ideas, but I have tried to sketch the major shifts in the popular understanding of maths. The book is structured by a combination of historical and thematic considerations, and its thirteen chapters can be grouped into four main sections. I begin by discussing the number concept, from a speculative and rhetorical account of prehistoric rituals to mathematics in the ancient world. I examine the relationship between counting and the continuum of measurement, and try to explain how the rise of algebra has dramatically changed our world.

The first section ends with 'mathematical padlocks' of the modern era, but in the second section I step back in time. More specifically, I discuss the origins of calculus, and the conceptual shift that accompanied the birth of non-Euclidean geometries. In short, I try to explain how modern mathematics grew beyond the science of the Greeks, the Arabs, or other ancient cultures.

In the third section I turn to the most philosophically loaded terms in mathematics: the concept of the infinite, and the fundamentals of formal logic. I also discuss the genius of Alan Turing, and try to elucidate the subtle relationship between truth, proof and computability. In particular, I focus on a proof of the infinite richness of addition and multiplication (as demonstrated by Matiyasevich's Theorem), and examine Kurt Gödel's celebrated theorems on the Incompleteness of Arithmetic.

In the final section I consider the role of mathematics in our attempts to comprehend the world around us. In particular, I describe the importance of models, and the role of mathematics in biology. I conclude by taking a step back from any particular theorem, and try to use what we have learned about mathematical activity to think about thinking in general.

One of the challenges in writing this book was doing justice to the weight of simple, teachable statements. Some statements are like paper darts: you can follow them with a lightness of contemplation, if you know to where they float. If your only guide is to cling to the words themselves, they cannot carry you, as their target has not been spoken. Other statements possess gravitas, as in their accessible simplicity they act like stones, pulling us down to what can and has been said.

Unfortunately, people tend to underestimate the value of simple, understandable statements, as we more often praise ideas by suggesting they are hard to grasp. As the great thinker Blaise Pascal remarked in *The Art of Persuasion*, 'One of the main reasons which puts people off the right way they have to follow is the concept they first encounter that good things are inaccessible by being labelled great, mighty, elevated, sublime. That ruins everything. I would like to call them lowly, commonplace, familiar. These names befit them better. I hate these pompous words ...'

The great edifice of mathematical theorems has a crystalline perfection, and it can seem far removed from the messy and contingent realities of the everyday world. Nevertheless, mathematics is a product of human culture, which has co-evolved with our attempts to comprehend the world. Rather than picturing mathematics as the study of 'abstract' objects, we can describe it as a poetry of patterns, in which our language brings about the truth that it proclaims. The idea that mathematicians bring about the truths that they proclaim may sound rather mysterious, but as a simple example, just think about the game of chess. By describing the rules we can call the game of chess into being, complete with truths that we did not think of when we first invented it. For example, whether or not anyone has ever actually played the game, we can prove that you cannot force a competent player into checkmate if the only pieces at your disposal are a king and a pair of knights. Chess is clearly a human invention, but this fact about chess must be true in any world where the rules of chess are the same, and we cannot imagine a world where we could not decide to keep our familiar rules in place.

Mathematical language and methodology present and represent structures that we can study, and those structures or patterns are as much a human invention as the game of chess. However, mathematics as a whole is much more than an arbitrary game, as the linguistic technologies that we have developed are genuinely fit for human purpose. For example, people (and other animals) mentally gather objects into groups, and we have found that the process of counting really does elucidate the plurality of those groups. Furthermore, the many different branches of mathematics are profoundly interconnected, to art, science and the rest of mathematics.

In short, mathematics is a language, and while we may be astounded that the universe is at all comprehensible, we should not be surprised that science is mathematical. Scientists need to be able to communicate their theories, and when we have a rule-governed understanding, the instructions that a student can follow draw out patterns or structures that the mathematician can then study. When you understand it properly, the purely mathematical is not a distant abstraction – it is as close as the sense that we make of the world: what is seen right there in front of us. In my view, maths is not abstract because it has to be, right from the word go. It actually begins with linguistic practice of the simplest and most sensible kind. We only pursue greater levels of abstraction because doing so is a necessary step in achieving the noble goals of modern mathematicians.

In particular, making our mathematical language more abstract means that our conclusions hold more generally, as when children realize that it makes no difference whether they are counting apples, pears or people. From generation to generation, people have found that numbers and other formal systems are deeply compelling: they can shape our imagination, and what is more, they can enable comprehension. The story of maths is fascinating in its own right, but in writing this book I hoped to do more than simply sketch a history of mathematical ideas. I am convinced that the history and philosophy of maths provide an invaluable perspective on human nature and the nature of facts, and I hope that my book conveys the subject's cultural, aesthetic and philosophical relevance, as well as the compelling drama of mathematical discovery.

Chapter 1:
BEGINNINGS

'There can be no doubt that all our knowledge begins with experience. ... But though all our knowledge begins with experience, it does not follow that it all arises out of experience.'

Immanuel Kant, 1724–1804

Language and Purpose

Researchers working with infants and animals have found compelling evidence that we have an innate sense of quantity. More specifically, humans, birds and many other animals can recognize when a small collection has changed in size, even if they do not observe the change taking place. For example, birds can recognize when one of their eggs is missing, even if they did not witness the egg's removal. Similarly, many animals will consistently pick the larger of two collections when they are given a choice. Presumably, this sensitivity to quantity is a necessary precondition for the development of maths, and it is interesting to note that some animals are quicker than humans at intuitively sensing differences in quantity. Nevertheless, although such abilities constitute evidence for animal intelligence, it is rather inaccurate to claim that 'birds count their eggs'.

I would argue that 'proto-mathematical' thinking can only begin once we have developed language, and that this kind of understanding is fundamental to many types of human behaviour, not just what we ordinarily think of as 'understanding maths'. Of course, any account of the lifestyle of our Stone Age ancestors is bound to be highly speculative, but despite the lack of conclusive evidence, I think it is helpful to imagine how our ancestors first developed rational capabilities, and the enormously complex thing that we call language.

Humans are not the only animals to use tools, and for millions of years our primate ancestors extended their abilities by utilizing what was found at hand. Sticks, stones, fur, leaves, bark and all manner of food stuffs were used in playful ways that we can only guess. Flesh was scraped from fur, sticks were sharpened and adapted to a purpose, and stones were knapped to produce effective butchery kits. Most importantly, about 1.8 million years ago *Homo erectus* started using fire to cook, which reduced the amount of energy needed for digestion, making it possible to grow larger brains and smaller digestive tracts.

As human intelligence evolved, our vocalizations and patterns of interaction developed into something that deserves to be called language. One very plausible speculation is that more intelligent hominids were more successful in making the most from the complex dynamics of their social situation, providing the selective advantage that led to increasing intelligence. In any case, communicative aspects of modern language are common to many animals. For example, many animals can convey a state of panic when they see a predator. It is therefore clear that complex, communicative forms of interaction massively predate the development of the proto-mathematical, or any conception that language might be the thing of interest, rather than the people who were making the sounds.

This idea is worth elaborating, so as an example of how a culture of interactions can evolve to something greater, imagine a woman who lives in a community with a particular culture of responses: *Men who give me tasty food will hear me hum, but those who grab me without giving me tasty food will hear me growl.* If a man was trying to establish a sexual relationship with this woman, he would want to hear the humming sound, because a woman who hums is much more likely to be interested in sex than a woman who growls. Consequently, the man would wilfully do a bit of cooking prior to any sexual advance, acting to establish the circumstances that he associates with the humming sound.

By living in such a social context, we came to feel the sense of our own actions. In other words, we responded to changing social occasions with increasingly sophisticated forms of motivated strategy, and were mindful of complex goals whose achievement required actions beyond those in the immediate present. For example, the occasion of preparing an especially tasty meal is not the same occasion as hearing a woman make the humming sound, but we see that one is motivated by the other.

Social norms and the vocabulary of praise and blame are both potent forces for shaping the imagination. It is absolutely fundamental that we find words with which to judge our actions, and our judgements work with words. An example of this endlessly subtle process can be found in the following conversation:

'Let's break into that house.'
'I don't know, that seems like a bad idea.'
'Go on, don't be chicken.'

We are fearful that our reasoning will compel us to name ourselves cowards, idiots, or many other kinds of undesired utterance. The will to avoid such experience is part of our humanity, as is the compelling nature of the reasons that we find. As Blaise Pascal observed, we are most compelled by the reasons of our own devising, but such complexities can be closely shared and instinctively taught to others. The caveman is compelled by the fact that the woman has established reasons for growling, if he fails to meet the expectations of an established practice. It is a process of judgement that he has a feeling for, and the weight of the utterance is that it is not felt to be arbitrary. Similarly, our potential thief is pulled by the fact that he too can reason himself a coward, and does not wish to do so.

However, it is crucial to note that in each of these examples, the significance of an utterance is inseparable from the fact that another person has decided to say the statement in question. In other words, we reached the point of very sophisticated communication long before we ever considered 'language' as a thing in itself, separate from the people who were speaking.

Human Cognition and the Meaning of Maths

The literature of mathematics is largely composed of arguments of the form 'If A and B are true, then it follows that C is also true', and it is worth pausing to wonder how it is that humans developed the capacity for deductive reasoning. We are not the only animals who are alert to the range of possible consequences of our actions, and we might suppose that our grasp of logical consequence is only possible because we have evolved the cognitive abilities needed to predict the practical consequences of the things that we might do. For example, imagine a hungry ape looking at another ape with some food. It might think to itself, 'If I grab the food, that big guy will hit me. I don't want to be hit, so therefore I should restrain myself and not grab the food.'

The fact that we use language fundamentally changes the character of our reasoning, but it is easy to believe that imagining the consequences of potential actions is an ancient ability that confers an evolutionary advantage. However, it is hard to see how the evolution of this kind of 'reasoning' about actions and their consequences could enable abstract thought. After all, the scenario I have described is all about judging the way to behave in a complex context, where any new information might change our prediction of what will happen next, and we ought to be open to noticing further clues. For example, if our ape saw the other ape make a friendly gesture, it might be wise to grab the food instead of letting it go. That is very different from working out logical consequences, where one thing follows from another, regardless of any further information that could plausibly come our way.

Because the social cunning of animals depends on their grasp of entire contexts, where there are always further clues, it is difficult to see how that kind of understanding could provide the cognitive abilities that a mathematician requires. In contrast, our capacity for spatial reasoning is much less open ended, and human beings do not need to be trained to make valid spatial deductions. For example, suppose that there is a jar inside my fridge. Now suppose that there is an olive inside the jar. Is the olive inside the fridge? The answer is yes, of course the olive is inside the fridge, because the olive is in the jar, and the jar is in the fridge. Now imagine that the jar is in the fridge but the olive is not in the fridge. Is the olive in the jar? Of course not, because the jar is in the fridge, and I have just told you that the olive is not in the fridge.

In reasoning about the location of the olive, it is sufficient to bear a thin skeleton of facts in mind. Additional information will not change our thinking, unless it contradicts the facts that form the basis of our deduction. Also note that in order to make these deductions, we do not need to be initiated into some or other method of symbolizing. All humans can reason in this way, so it is plausible to claim that there are innate neural mechanisms that underpin our grasp of the logic of containers. Of course, in order to pose these questions I need to use some words, but humans (and other animals) find it very easy to understand that containers have an inside and an outside, and this kind of understanding provides a structure to our perceptual world.

There is strong empirical evidence that before they learn to speak, and long before they learn mathematics, children start to structure their perceptual world. For example, a child might play with some eggs by putting them in a bowl, and they have some sense that this collection of eggs is in a different spatial region to the things that are outside the bowl. This kind of spatial understanding is a basic cognitive ability, and we do not need symbols to begin to appreciate the sense that we can make of moving something into or out of a container. Furthermore, we can see in an instant the difference between collections containing one, two, three or four eggs. These cognitive capacities enable us to see that when we add an egg to our bowl (moving it from outside to inside), the collection somehow changes, and likewise, taking an egg out of the bowl changes the collection. Even when we have a bowl of sugar, where we cannot see how many grains there might be, small children have some kind of understanding of the process of adding sugar to a bowl, or taking some sugar away. That is to say, we can recognize particular acts of adding sugar to a bowl as being examples of someone 'adding something to a bowl', so the word 'adding' has some grounding in physical experience.

Of course, adding sugar to my cup of tea is not an example of mathematical addition. My point is that our innate cognitive capabilities provide a foundation for our notions of containers, of collections of things, and of adding or taking away from those collections. Furthermore, when we teach the more sophisticated, abstract concepts of addition and subtraction (which are certainly not innate), we do so by referring to those more basic, physically grounded forms of understanding. When we use pen and paper to do some sums we do not literally add objects to a collection, but it is no coincidence that we use the same words for both mathematical addition and the physical case where we literally move some objects. After all, even the greatest of mathematicians first understood mathematical addition by hearing things like 'If you have two apples in a basket and you add three more, how many do you have?'

As the cognitive scientists George Lakoff and Rafael Núñez argue in their thought-provoking and controversial book *Where Mathematics Comes From*, our understanding of mathematical symbols is rooted in our cognitive capabilities. In particular, we have some innate understanding of spatial relations, and we have the ability to construct 'conceptual metaphors', where we understand an idea or conceptual domain by employing the language and patterns of thought that were first developed in some other domain. The use of conceptual metaphor is something that is common to all forms of understanding, and as such it is not characteristic of mathematics in particular. That is simply to say, I take it for granted that new ideas do not descend from on high: they must relate to what we already know, as physically embodied human beings, and we explain new concepts by talking about how they are akin to some other, familiar concept.

Conceptual mappings from one thing to another are fundamental to human understanding, not least because they allow us to reason about unfamiliar or abstract things by using the inferential structure of things that are deeply familiar. For example, when we are asked to think about adding the numbers two and three, we know that this operation is like adding three apples to a basket that already contains two apples, and it is also like taking two steps followed by three steps. Of course, whether we are imagining moving apples into a basket or thinking about an abstract form of addition, we don't actually need to move any objects. Furthermore, we understand that the touch and smell of apples are not part of the facts of addition, as the concepts involved are very general, and can be applied to all manner of situations. Nevertheless, we understand that when we are adding two numbers, the meaning of the symbols entitles us to think in terms of concrete, physical cases, though we are not obliged to do so. Indeed, it may well be true to say that our minds and brains are capable of forming abstract number concepts because we are capable of thinking about particular, concrete cases.

Mathematical reasoning involves rules and definitions, and the fact that computers can add correctly demonstrates that you don't even need to have a brain to correctly employ a specific, notational system. In other words, in a very limited way we can 'do mathematics' without needing to reflect on the significance or meaning of our symbols. However, mathematics isn't only about the proper, rule-governed use of symbols: it is about *ideas* that can be expressed by the rule-governed use of symbols, and it seems that many mathematical ideas are deeply rooted in the structure of the world that we perceive.

Stone Age Rituals and Autonomous Symbols

Mathematicians are interested in ideas, not just the manipulation of 'meaningless' symbols, but the practice of mathematics has always involved the systematic use of symbols. Mathematical symbols do not merely express mathematical ideas: they make mathematics possible. Furthermore, even the greatest mathematicians need to be taught the rules before they can make a contribution of their own. Indeed, the very word mathematics is derived from the Greek for 'teachable knowledge'. The question is, how and why did human cultures develop a system of rules for the use of symbols, and how did those symbols change our lives?

It seems fair to claim that the most fundamental and distinctive feature of human cognition is our boundless imagination. We don't just consider our current situation, we imagine various ways that the future could pan out, and we think about the past and how it could have been different. In general, we inhabit imaginable worlds that follow certain principles, and compared to other animals, our thoughts are not overly constrained by our current situation or perceptions. In particular, we can think about objects that are not ready at hand, and it is reasonable to assume that in the distant past, our ancestors would feel distressed if their desire to act was frustrated by the absence of some object or tool.

As an animal might express the presence of predators, our ancestors would gesture, 'I am missing a flint'. By using their vocal cords, facial expressions and bodily posture, they would express their motivated looking. Fellow primates would respond to this signal in a manner appropriate to the occasion, having an empathic grasp of what it is to search in such a fashion. Over countless generations, our ancestors must have developed ways of conveying a desire for certain objects, even though those objects were currently out of sight. Furthermore, at some point our ancestors must have made the vital step of imbuing those expressive gestures with an essentially mathematical meaning. This remarkable feat was not achieved by the discovery of abstract objects: it was achieved by developing rituals.

Suppose, for example, that there was a pre-existing form of expression that conveyed the speaker's irritation over missing a flint. Now imagine the earliest people running their hands over their treasured tool kit of flints. As a person checked their tools time and time again, they may have expressed their familiarity with these objects by reciting a sequence of names. As each tool was touched in turn, we can imagine our ancestors repeating a distinctive sequence of rhythmic speech, with one word for each tool, like someone saying 'Eeny-meeny-miney-mo'. If this ritual was left unfinished by the time there were no more objects left to touch, Stone Age humans could see that they had a *reason* for making the gesture 'I am missing a flint'.

This is not the same as counting with an abstract concept of number, as whether or not they deserve to be called mathematicians, even the youngest of children will not mistake 'Eeny-meeny-miney' for 'Eeny-meeny-miney-mo'. Sensitivity to the incompleteness of a habitual action is clearly innate, and this is very close to the sing-song voice of baby talk and our natural sense of rhythm. When our ancestors expressed this failure of correspondence between the present tools and the familiar ritual, the other people would also know that something was missing, because they recognized that the ritual had been performed correctly. In other words, it is the *ritual* that tells us that a flint is missing, and not just the individual who performed it.

By possessing such a clear sense of justified speech, people could find grounded meanings in their utterances, and strategically approach the issues that concerned them (namely, 'Is it the case that all the familiar devices are present?'). In this way, expressive gesture came to signify more than an immediate cultural resonance, and the primal gestures that conveyed the sense 'I am irate over a missing flint' become something deeper. The common sense of valid reasoning gave new weight to our communications, as by means of the common practice a statement of fact can be established. In particular, note that the fact's appearance in the world is dependent on the practice itself, not the individual who carried it out. This process of language speaking for itself emphatically does not require the abstract concept of number, and I would argue that proto-mathematical thought had a gradual evolution that predates counting by many tens of thousands of years.

The origin of number words as we understand them today isn't known for certain, but there are some interesting theories supported by linguistic evidence. It may be that practices closely related to counting arose spontaneously throughout the world, more or less independently from place to place. However, the mathematician and historian of science Abraham Seidenberg (1916–1988) proposed that counting was invented just once, and then spread across the globe. Number words are often related to words for body parts, and Seidenberg claimed that the similarities in number words from very distant places constitutes evidence for his theory. He also made the intriguing observation that in almost every numerate culture, there is an ancient association between the odd numbers and the male, while the even numbers are female.

There is certainly plenty of evidence that animals are aware of who is first in the pecking order, who is second, third and so on. Seidenberg suggests that counting originated in rituals based on rank and priority, arguing that counting 'was frequently the central feature of a rite, and that participants in the rite were numbered'. Whether the first numbers or number-like words were applied to an ordered sequence of people, or used to assess the plurality of a collection of tools, it is clear that the human mind has been capable of learning how to count for tens of thousands of years.

It is important to note that mathematics is not a universal human trait, as some cultures have no words for numbers larger than three. Furthermore, some people have a highly cultured sense of quantity even though they cannot really count, as their language has too few number words.

For example, the Vedda tribesman of Sri Lanka used to 'count' coconuts by gathering one twig for each coconut. The people who did this clearly understood that there was a corresponding plurality between the twigs and the coconuts, but if asked how many coconuts a person had collected, they could only point to the pile of twigs and say 'that many'.

The first expressions of quantity are lost deep in the mists of time, but it is surely safe to assume that long before the advent of abstract number words, people had one word for 'hand', and a different word for 'pair of hands'. The move from words that convey quantities of specific physical things to an abstract or universally applicable language of number is an example of logic at work. That is to say, once we have a sequence of words for 'counting' something or other, it is possible to recognize that it is the words themselves that form an ordered, rule-governed sequence, and they need not be bound to any particular thing that people are used to counting.

Making Legible Patterns

As human beings we live in a world of people and things, sights and sounds, tastes and touches. We don't see a pattern on our retina: we see people, trees, windows, cars, and other things of human interest. This relates to the fact that we use language to think about our world, doing things like naming objects, or creating accounts of people or situations. My point is that human beings conceptually structure the perceptual flux in which we live, so our use of symbols, images and words is central to making sense.

For example, imagine a young child drawing a picture of Daddy: a stick-man body with a circle for a head, two dots for eyes and a U-shape for a smile. It is significant that each part of this drawing can be named, as we understand, for example, that two dots can represent the eyes. The art of 30,000 BC was probably somewhat similar to a child's drawing, not because our ancestors were simple minded, but because drawing nameable things is such a basic, human skill. Indeed, we can say that children's drawings are understandable precisely because we can talk our way about them.

Our ancestors decorated caves with vivid illustrations of large mammals, but they also used simpler marks (arrangements of dots, V-shapes, hand prints, etc.). Just as a child might not need to draw ears and a nose before their marks become a face, so the caveman artist may have drawn some tusks and already seen a mammoth. Such stylized, intelligible drawings are not the same as writing, but there is a related logic of meaningful marks, and it is surely safe to assume that our ancestors talked about their drawings. As another example of Stone Age pattern making, archaeologists in central Europe found a shinbone of a wolf marked with fifty-seven deeply cut notches. These marks were arranged in groups of five, and carbon dating indicates that this bone is over 30,000 years old.

People and animals alike are good at spotting patterns. In particular, many birds and mammals are demonstrably sensitive to changes in quantity. Time and time again humans have discovered a basic technique for clearly showing quantity: we group elements together in a regular way, so that a single 'counting' operation is broken into a combination of simpler assessments. For example, we can recognize four as a pair of pairs, or ten as a pair of fives. This means that even before one can count, it is easier to assess the plurality of things if they are arranged in regular groups, rather than scattered in a disordered fashion. Furthermore, once we have words for numbers, this practical idea can lead us to the concept of multiplication. This suggests that by the time that our ancestors had articulated an abstract concept of number, people were counting by dividing their things into regular groups and counting off five, ten, fifteen, twenty (say). In other words, the times tables may be just as old as abstract number words themselves.

Mathematics has been described as the language of patterns, and there is a deep relationship between our innate tendency to recognize patterns, and our cultured sense of shape and number. Long before we developed proper number words, ancient peoples must have recognized the reality of patterns, and explored some formal constraints. Very ancient peoples must have known that triangles can be arranged to produce certain shapes or patterns, but there are some shapes (e.g. a circle) that cannot be made from triangles. People have been exploring patterns for tens of thousands of years, using their material ingenuity (e.g. pottery and basket weaving), music, dance and early verbal art forms. For example, it is an evident truth that if you clap your hands every second heartbeat, and stamp your feet every third one, there necessarily follows one particular combined rhythm and not others.

The meaningfulness of mathematical statements did not appear from nowhere, and we don't need 'proper' mathematics to first be aware of quantity and shape. Before people developed counting or abstract number words, they might have used a phrase like 'as many bison as there are berries on a bush', or shown a quantity with an artistic abundance of marks. After all, our artist ancestors surely strove to be eloquent, and someone must have been keen to show the size of a massive herd. Many generations later, the advent of counting gave birth to the concept of number: a great advance in our ability to conceptually distinguish between different pluralities.

At first our ancestors must have only used their counting

words in particular situations, but over time we realized that in order to count any collection of objects, we do not need to keep our eyes on the qualities of the objects themselves. In a sense we can count any collection of objects, so long as we can give each object a name (e.g. by attaching labels). From that point on we can play the counting game simply by reviewing the sequence of names that we ourselves have given, even if our labels become detached from their associated objects.

Many, many social needs require calculation and number, and over the long arc of prehistory mathematics continued to evolve along with the social systems that supported mathematical techniques. In return, more sophisticated mathematics enabled more complex social structures. For example, an inheritance cannot be distributed unless certain facts about division are known, and at a more sophisticated level, tax rates and monetary systems are literally inconceivable without the concept of number.

The development of agriculture revolutionized our ways of life, and according to many ancient historians, geometry (Greek for earth-measuring) came into being as people needed to speak authoritatively and uncontentiously about the size of fields. In particular, every year Egyptian mathematicians needed to replace the property markers that were washed away from the flood plains of the Nile. The story of geometric techniques arising with the need to measure fields is certainly plausible, but there were also prehistoric traditions for communicating specific plans for temples and other buildings, which necessarily involves a language of shape, so this claim may not actually be correct. What is known for certain is that by the third millennium BC, civilizations with sophisticated mathematical practices had developed along the fertile banks of many of the world's great rivers. The Nile, the Tigris and Euphrates, the Indus, the Ganges, the Huang He and the Yangtze all provided ground for these new ways of life. Furthermore, mathematics had a central role to play in the emergence of large-scale civilizations, not least in the development of trade, measured and planned construction, administrative techniques, astronomy and time keeping.

The Storage of Facts

You might guess that the oldest recorded numbers would be fairly small, but in fact archaeologists have found that some of the oldest unambiguously numerical records refer to the many thousands of cattle that Mesopotamian kings had claimed through war. It is also interesting to note that no civilization has ever become literate without first becoming numerate, and almost every numerate civilization is known to have used some kind of counting board or abacus. In other words, tools for recording the counting process are much, much older than tools for recording speech. Greeks and Romans used loose counters, the Chinese had sliding balls on bamboo rods, and the Ancient Hindu mathematicians used dust boards, with erasable marks written in sand.

Because of geographical distance, it is assumed that the development of mathematics in the Americas was completely independent from that of Europe and Asia. It is therefore remarkable to note that around 1,500 years ago, the Maya were employing number symbols much as we do today. The Incas were a more recent civilization than the Maya, and although the Incas never developed a system for making records of the spoken word, they could record information by using a system of knotted cords called *quipus*. These were colour coded to represent the various things that were counted, and scribes would read the clusters of knots by pulling the cord through their hands. Each cluster of knots represented a digit from one to nine, and a zero was represented by a particularly large gap between clusters.

Quipus with as many as 1,800 cords have been found, and as different-coloured threads signified different kinds of information, these fascinating objects demonstrate that sophisticated record keeping is not the exclusive domain of the written word. Very few quipus have survived, and given that they are far more distant than the Incas, it is plausible that many prehistoric civilizations possessed long-lost means of embodying data. It is worth remembering that only a tiny subset of equipment survives the ravages of time, and if people kept records by arranging pebbles or making scratches on bark, we might have no way of knowing. As the mathematical historian Dirk Struik has suggested, the builders of ancient monuments like Stonehenge must have had some idea of what it was that they were building. Many of the regular features of this construction cannot be accidental, and it seems highly unlikely that the builders didn't know what they would do with the stones until they got them on site. Their means of communicating intent may well have involved physical artefacts that embodied data unambiguously. For example, they may have made shadow casting models that showed exactly how many stones were going to be arranged, and their orientation in relation to the sun's path.

The ancient civilizations of Asia used bamboo, bark and eventually paper to keep records of numerical information. Although the origins of their mathematical knowledge remain obscure, certain pieces of Chinese mathematics have been faithfully passed down through hundreds of generations. For example, consider the following 'magic square', known as the *Lo-Shu*. Legend has it that this mathematical pattern emerged from the Yellow River on the back of a giant turtle about 4,000 years ago. We can't really be certain about the true age of the *Lo-Shu*, but we do know that it was considered to be truly ancient knowledge as far back as the Han dynasty (206 BC–AD 220).

4	9	2
3	5	7
8	1	6

Every row, column and diagonal contains numbers that add up to fifteen. The central cross of odd numbers ensures that every line contains either one odd number or three odd numbers (the 'yang' of the *Lo-Shu*). The four corners contain even numbers (the 'yin' of the *Lo-Shu*), completing this sacred symbol of cosmic harmony and balance.

Perhaps the most famous book of Ancient Chinese mathematics is the *Nine Chapters on the Art of Calculation*. This book was written *c.* 200 BC, and the 246 problems it contained were used to test and train potential civil servants. Arguably the most impressive feature of this mathematical tradition is the fact that the Ancient Chinese routinely solved systems of linear equations. For example, suppose we have two different kinds of weights, coloured red and blue. If two reds and three blues weigh eighteen units, while two reds and two blues weigh sixteen units, how much does one red weigh? In modern notation we have:

$$2r + 3b = 18 \text{ and } 2r + 2b = 16.$$

The left-hand side of the first equation minus the left-hand side of the second is equal to b. The right-hand side of the first equation minus the right-hand side of the second is equal to 2. Therefore b, the weight of one blue is 2 units. Substituting this value into either equation shows that $r = 6$. Modern mathematicians and the Ancient Chinese alike solve this kind of problem by adding and subtracting equations (or equation-like forms) in order to eliminate unknowns, and the same principle can be naturally extended to situations with more than two unknowns. The Chinese have been solving problems like this for over 2,000 years, but remarkably, this extremely useful technique was not known in the West until the beginning of the nineteenth century, when it was independently invented by one of the giants of modern mathematics, Carl Freidrich Gauss.

Babylon, Egypt and Greece

Around 1650 BC, a scribe named Ahmes copied out a text from the Twelfth Dynasty of Egypt (*c.* 1990–1780 BC). It seems that Ahmes was mathematically educated, which makes him the most ancient mathematician whose name we know. The 'Ahmes Papyrus' contains eighty-five problems, and it demonstrates the Egyptians' ability to solve problems that involve unknown quantities, as well as their systematic use of fractions of the form $^1/_n$. For example, the 'Ahmes Papyrus' contains the line 'a heap plus a quarter of that heap again makes fifteen'. Through trial and error Ahmes realized that the heap must be size twelve (because twelve plus a quarter of twelve is fifteen). The manuscript also states that the area of a circle of diameter nine can be taken to be equal to the area of a square of width eight. This represents an error in the estimate of π of only 0.6%.

Although many modern teenagers know significantly more mathematics than any Ancient Egyptian, it is clear that the Egyptians knew how to achieve a fairly rich variety of computational tasks. In particular, they had a long tradition of measuring the position of the stars and planets, and after the annual flooding of the Nile they were more than capable of replacing the property markers that had been washed away. A particularly ancient measurement technique involves a form of set square, made from a loop of rope marked into twelve equal lengths. The Egyptians and other ancient peoples knew that if you stretched out a rope with one side of 3 units, one side of 4 units and one side of 5 units, you invariably produced a right-angled triangle. Indeed, the word 'hypotenuse' is derived from the Greek for 'stretched against', reflecting this ancient technology. The Egyptians also knew that $3^2 + 4^2 = 5^2$, and like other ancient peoples they were familiar with many other Pythagorean triangles.

The mathematicians of Mesopotamian were probably more advanced than their Egyptian contemporaries. By the time that Hammurabi became king of Babylon (*c.* 1750 BC), his people had developed powerful methods for finding areas and volumes. Indeed, the Babylonians were familiar with the empirical content of Pythagoras' Theorem over a thousand years before Pythagoras himself was born. Our main sources of evidence concerning Babylonian mathematics are the many clay tablets that have survived, preserving the maths homework of young scribes from

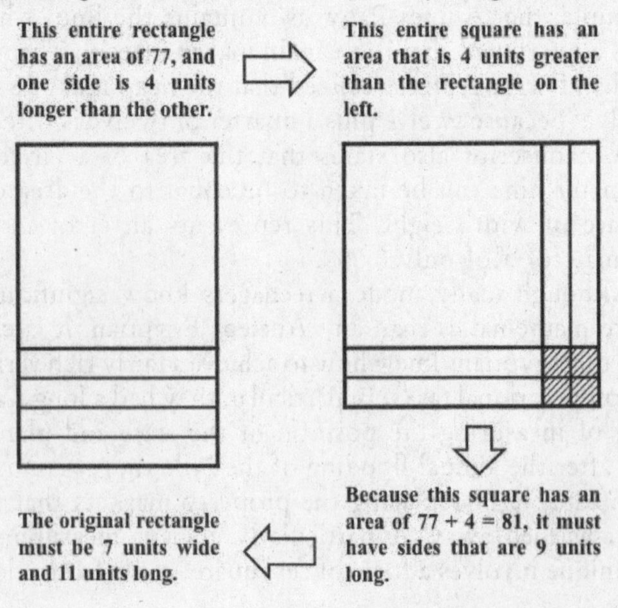

over 3,500 years ago. Like the Egyptians, the Babylonians would sometimes phrase questions in terms of unknown quantities, and remarkably, they knew effective methods for finding positive solutions to quadratic equations. More specifically, they would use their knowledge of the area of squares and rectangles to answer questions like the following: 'A rectangular field has an area of $77m^2$ and one side is $4m$ longer than the other. How long are the two sides?'

Mesopotamian mathematics co-evolved with a desire to study the night sky (a feature of the natural world that had great religious significance for many ancient peoples), and it also co-evolved with the practical tasks of taxation, trade and measurement. Without a mathematical language, tax records could not have been kept, nor would it have been possible to record accurately the passage of the stars. Mathematics enabled the bureaucratic procedures that are required to rule a large state and, conversely, the novel problems that confronted the first cities must have provided a potent breeding ground for newly elaborated forms of mathematics.

The range of ancient problems and exactitude of some of the surviving calculations strongly suggests that symbolic calculation was deeply valued, and cultivated as a broadly applicable skill, transcending its role in any concrete, practical task. Pythagoras' Theorem in particular has been applied to a huge range of both practical and poetic problems, and it is clear that ancient peoples relished the fact that the same numerical relationship holds whether we are talking about the length of fields, spears, shadows or buildings. For example, an Ancient Egyptian manuscript asks 'If a ladder of 10 cubits has its foot 6 cubits from the wall, how high does the ladder reach?', and this is clearly not a question that needed to be answered for pressing practical reasons.

It has often been argued that the oldest mathematical procedures were simply something that you did: a kind of ritual of computation that did not stand in need of justification. For example, the Ancient Egyptians were acutely concerned with the concept of Ma'at. This concept was personified as a goddess, whose name can be translated as truth, order or justice. The natural world, the state and the individual were all parts of the domain of Ma'at, and for several millennia, rulers would flatter themselves by being described as 'Lords of Ma'at', or preservers of the divine order.

For the Ancient Egyptians, following the time-honoured ways of their ancestors was literally believed to be a matter of life and death. The divine order and cosmic harmony of Ma'at could turn to chaos and violence if the ruler or his people did not adhere to their traditions and rituals. To put it bluntly, Ancient Egyptians believed that they had a sacred duty to do things as they had 'always' been done, and a priest who was writing mathematics would not have thought it a good idea to invent a new way of doing things. After all, if a time-honoured way with symbols remains eloquent, why argue with your teachers? In particular, people would be shocked and angry if you suggested changing the way that property markers were replaced after the flooding of the Nile, as people's confidence in the process quite rationally depended on the fact that it was done the same way every year.

In contrast, later generations of Greek historians were justifiably proud of the fact that in Greece learned men *debated* mathematical truths, and actively tried to develop new forms of mathematics. Instead of merely having a ritual of computation, they engaged in arguments, and tried to deduce an expanded vision of the truth. Of course, earlier cultures had also made mathematical deductions. The difference is that when people in earlier cultures were initiated into mathematical practices, they were taught through the presentation of examples. Teachers may well have made quite abstract, elucidatory remarks as to why particular techniques were effective, but the emphasis on mathematical *proof*, and the rigorous articulation of logical principles, are characteristically Greek innovations.

The Logic of Circles

As the Greeks debated, they articulated well-defined conceptual schemes, and deduced a range of very general truths. This process led to a significantly more abstract form of mathematics, as Greek thinkers made a point of stressing that they were interested in conceptual principles, and not simply working with particular collections of actual, countable things, or actual measured objects. Their radical innovation was to construct arguments that definitively settled the facts of various mathematical matters, and their geometric deductions were made in the presence of describable, labelled diagrams.

This final point is very important. As we shall see in later chapters, modern mathematical arguments often hinge on a kind of calculation that essentially rests on the definitive properties of *symbols*. Ancient Greek mathematics was somewhat different. It did involve abstract symbolism (e.g. a diagram might include an arbitrary length 'AB' or an angle 'ABC'), but the mathematical objects they discussed were all idealizations that were not too far removed from our experience with actual objects. For example, in classical geometry a curve is said to be 'a path traced by a moving point'. The movement involved is understood to be metaphorical, as a circle (for example) does not literally need to appear over time, drawn by a point moving at some or other speed. We simply describe what is meant by 'a curve' by talking in terms of motion, because we all know what it is like to follow a path with the mind's eye. Indeed, this metaphor of motion is also found in non-mathematical language, as when we say 'the road runs through the wood', or 'the road goes over the hill', we do not mean to imply that the road is literally moving.

Similarly, in classical geometry the term 'straight line' refers to something abstract, but the meaning of the word is nevertheless grounded in experience. In particular, it seems fair to say that our grasp of the concept of straight lines is partly informed by our experience with taut strings. Of course, this doesn't mean that our idea of a straight line is merely some mental image of a taut string, as even the finest string has some thickness, while a 'straight line' is an abstract, mathematical form that by definition has no thickness at all. For many centuries European mathematics was completely dominated by geometric concepts of this sort, and it is worth pausing for a moment to remember our initiation into the language of shape. As children we are taught to recognize named shapes, and we quickly take this game for granted. However, drawing a loop in the sand and naming it 'circle' is a truly remarkable act. Almost every aspect of the drawing is irrelevant, because by seeing the drawing *as* a circle (that is, by seeing the rightness of our description), we encounter the very sense that the geometer is interested in.

To look for such forms is to engage in the process of abstraction: trying to see the thing in its most elemental of characteristics, and no more. We simply do not and should not care whether the circle is perfectly drawn, or whether it has been traced in sand or scratched on wood. This is in marked contrast to something like a portrait, where there is always more to talk about than the name of the person painted. The spherical symmetry of sun and moon suggests that mankind has been cognisant of circularity for a very long time indeed, having the capacity to draw the attention of others to the visual property of roundness. Of course, it is comparatively easy to learn what a circle looks like. More difficult is the task of stating the defining characteristic, namely that all the points on the edge are equally distanced from the centre.

Identifying the essential character of circles enables comprehension. For example, the fact that circles have this defining centre explains the shape of ripples in a pond. Before the stone hits, the water's surface is flat, and every point on the pond is very much like all the others. The impact of the stone disrupts this equilibrium, and points that are an equal distance from the disruption experience the same effects (rising and falling at the same time). In other words, because the stone marks a centre and distance is the relevant feature (not the particular direction from the centre), circular ripples are the evident consequence of dropping a stone in a pond. Similarly, if we ignore the effects produced by rotation, a liquid planet will form a sphere, so that each point on the surface is equally close to the centre of gravity.

The Factuality of Maths

We have seen that the Ancient Babylonians could find the length of a rectangular field, given the area of the field, and the difference between its length and breadth. If these people only had a narrow, practical interest in the length of fields, they could have simply measured them. After all, in what bizarre circumstance does a person know the difference between a field's length and breadth, without actually knowing either the length or the breadth? As this example indicates, it is clear that our ancestors' passion for mathematics ran deeper than narrowly practical concerns.

People are naturally drawn into investigating the world that our language evokes, and in many ways the mathematics of the ancient world is still accessible to us today. However, even though we can replicate the procedures of ancient mathematicians, our attitude towards mathematic facts has become very different. Up until the nineteenth century mathematics was considered to be a realm of 'hard facts'. Mathematical axioms were taken to be absolutely true, and any valid deductions based on those axioms were accepted as being facts about the universe. In other words, up until the nineteenth century mathematics could be accurately described as the science of shape and number, and nobody had thought to divide the subject into 'pure' and 'applied'.

The facts of mathematics are different from the facts of physics, but when we talk about straight lines, our words are intelligible because (among other things) we have encountered 'lines' in the real world. There is a long tradition of somewhat mystical rhetoric about how mathematical lines are 'perfectly' straight, unlike the pieces of string we might use to measure a field. Nevertheless, mathematical truths are of a piece with the only world we can know, even if they are further abstracted from experience than most of the truths we are interested in. After all, every statement that humans are capable of understanding must have some kind of grounding in our cognitive capabilities and our experience of the world, even if our interest is in the abstract concept rather than its particular instantiations.

Many people are rather confused about the relationship between physical reality and mathematical truth, as they forget that physical reality is not the same thing as our conception of it. For example, it is sometimes said that 'mathematics is true in every possible world', but that is only true in the sense that in every *conceivable* world you cannot force checkmate against a lone king with only a king and a pair of knights. Similarly, it is sometimes said that 'the universe runs according to mathematical laws', but that does not mean we ought to accept the quasi-religious belief that mathematics itself somehow forces the world to behave as it does. For example, it is a mathematical truth that the circumference of a circle is π times its diameter, but that does not mean that the number π is literally present in every circular object.

The empirical fact is that if we measure a circular object we need a piece of string a little over three times as long as the string we need to measure its diameter, but there is no empirical proof that the ratio between the two is precisely π. In fact, quantum theory tells us that there is a fundamental limit to the accuracy of any measurement of length, so physical lengths can never determine the infinitely precise quantities of mathematics. My point is that a circle is a concept, not something physical, and there is no empirical evidence to support the claim that the physical world is literally forced to obey the laws of mathematics. Of course, *physics* is deeply mathematical, as we use mathematical concepts to describe the regularities that are empirically observed. In other words, the fundamental fact is the regularity of the universe, but mathematics only enters the picture when we try to comprehend, describe and explain those regularities.

We cannot attempt to understand the world without using concepts, and although the gulf between mathematical and non-mathematical concepts is sometimes overstated, from the very beginning people recognized that mathematical truths have a distinctive quality. In particular, number concepts are very abstract indeed, and we have a strong sense that when we are talking about numbers we are not merely talking about the actual things we happen to be counting. The equivalence between counting a list of names and counting the named objects means that as mathematicians we are free to disregard whatever has been counted, so as children learn quite quickly, it really doesn't matter whether you are counting apples, pears or people. The abstractness of number concepts is very significant, but we sometimes forget that many other everyday words are similarly abstract. For example, we talk about the world in terms of colour, and have the related concepts 'same' colour and 'different' colour. Talking this way suggests the idea of a universal colour chart, though of course we should take account of the fact that our perception of colour is highly sensitive to context, and entirely dependent on light conditions.

Any particular 'universal colour chart' will be a rather arbitrary affair, as we can pick any degree of accuracy to qualify as 'the same colour', which means that we can squabble interminably over how many different colours should be on our chart. The situation is much more satisfactory in the case of numbers, as the way we count in itself generates the representatives that we require. The obvious possibility of translation means that it doesn't really matter whether we count in English or in French, using fingers or some beads. It is very important that we can use mathematical vocabulary in the marketplace or other specific contexts. However, the facts about our mathematical words come from the mathematical framework to which those words belong, and not from any wider sense of meaning that such words might gain by being practically employed.

As we shall see, mathematicians in the nineteenth century developed non-Euclidean geometries. This imaginative feat subtly altered our basic conception of maths, as only modern mathematicians would say, 'Let's just assume that these axioms are true, and then talk about the things we can deduce.' The art of proper, symbol-based reasoning has always been central to mathematics. What changed in the nineteenth century is that instead of starting with a hard fact, mathematicians now felt free to begin with 'mere' statements, which did not need to be 'actually true'. To put it another way, over the centuries the idea has developed that scientists study the real, physical world, while mathematicians study abstract objects that might or might not have anything to do with the real world.

This subtle shift in the philosophical roots of mathematics led to the invention of 'pure' and 'applied' maths. It is often assumed that these are fundamental, natural categories, but until the late nineteenth century, no mathematician would have known what you were talking about if you asked whether they studied pure or applied mathematics. Of course, there have always been practical problems that require mathematical expertise. Indeed, most great mathematicians have made a significant contribution to science, and to this day, many of the most compelling mathematical challenges directly relate to the languages that we use when describing physical phenomena. Even though such concerns are not a part of mathematics itself, the problems that we are interested in solving have always exerted a powerful influence on the development of maths.

The fundamental point is that theory and abstract language are inescapable parts of our attempts to describe the world. After all, we cannot present an account of the world without a language that expresses our theory. Furthermore, in the case of scientific theories, it is imperative that our language is capable of supporting precise, unambiguous deductions. We don't just want to say 'this is what the world is like'; we want to say 'because this is what the world is like, it follows that ...' Over time, scientists and other theorists have refined our definitions, and as we clarify the logical import of our statements, the disciplines of science have moved closer and closer to mathematics. As the uses of mathematics have changed, our sense of what the subject is about has also shifted, but ever since the Ancient Greeks, we have appreciated that mathematical concepts have a certain autonomy, as all mathematics is logical and essentially systematic.

Chapter 2: FROM GREECE TO ROME

'In arithmetic we are not concerned with objects which we come to know as something alien from without through the medium of the senses, but with objects given directly to our reason and, as its nearest kin, utterly transparent to it.'

Gottlob Frege, 1848–1925

Early Greek Mathematics

We don't know much about Greek mathematical practice in the pre-archaic period (before 650 BC), but we do know that the civilizations in that part of the world were all numerate, they were familiar with the concept of measured distances, and they employed trained scribes to keep numerical records. The negotiations of trade and the invention of money must also have encouraged numeracy and stimulated mathematical development. Indeed, the legendary father of Greek mathematics Thales of Miletus was said to have been a merchant.

With the older figures of Ancient Greece it is nigh-on impossible to separate fact from fiction, but it is thought that Thales visited Egypt and Babylon in the sixth century BC, learning something of their mathematical skills and engaging in learned debate. Some people's methodologies were seriously flawed, and distant mathematicians sometimes disagreed with one another. For example, the Egyptians were confident of their rule for finding the volume of a truncated pyramid, but the Babylonians had a conflicting one. It was clear to Thales that only one of them could be correct (in this case it was the Egyptians), and the kind of argument that he used to settle the fact of the matter could be considered the earliest form of modern, mathematical proof.

The crucial point is that the Ancient Greeks were interested in developing a theory of mathematics, and the construction of meta-narratives was characteristic of Greek scholarship. They didn't just write about the behaviour of good or bad rulers, they wrote about the theory of politics. Likewise, they didn't just write about techniques for healing, they wrote about the theory of medicine. It is also important to note that for the Ancient Greeks (and many later thinkers), the capacity for rational thought was considered to be one of the cardinal virtues of humanity. Mathematics provided an ideal arena in which to refine our logical abilities, and by the end of the fifth century BC the basic principles of mathematical deduction were firmly established.

For example, around 440 BC the Ionian philosopher Hippocrates of Chios wrote about the area of a crescent shape, and in doing so constructed arguments that logically led the reader from truth to truth. In other words, he didn't simply tell people how to calculate the given areas, he showed that his answers were logical consequences of an explicit list of axioms. Not all Greek mathematics was of the form 'by definition statement A is true, and if A is true B is true, and if B is true C is true ...', but the Greeks were certainly keen to identify the starting points of mathematical arguments, and their axiomatic method made a powerful impression on countless generations of thinkers.

Before this time, people had said things like 'to calculate this, follow such and such a procedure'. What was new about Greek mathematics was the fact that the entire means of deduction were rigorously stated, using diagrams and writing to bring the obvious to light. Indeed, the mathematician's well-justified aversion to unstated assumptions is one of the great legacies of the Ancient Greeks. They did not simply present a pattern and then say true things about it: they defined their terms, and then used their definitions to support deductive practice.

A system of rigid definitions is a surprisingly powerful thing, and the Greeks could employ their *logos* or language to remarkable effect. For example, consider the proof of Pythagoras' Theorem presented in the introduction. By explicitly stating all the relevant facts, we can *deduce* Pythagoras' Theorem from the statements 'squares have four equal sides and four identical corners', 'moving a shape does not change its area', and '$a + b = b + a$'. Other ancient cultures did not develop the same interest in elucidating the way that one statement follows from another, but in Greek mathematics the logical relationship between true statements are the main focus of attention. Consequently, the simple and obvious statements that earlier people may have taken for granted acquired a new status. Henceforth such 'axioms' were not only considered to be true, they were seen as the foundations of an entire discipline.

Pythagorean Science

As the founder of a mystical tradition, Pythagoras is shrouded by legend. We know that he was born on the island of Samos around 580 BC, and that he was at least eighty years old when he died. It is said that he met with Thales, who encouraged him to visit Egypt and Babylon, to learn the things they had to teach there. Many of the Greek schools emphasized the reality of change, but Pythagoras and his followers looked to identify the eternal features of the world, particularly those that related to number. In search of such principles, Pythagorean scholars contemplated geometry, arithmetic, astronomy and music, guided by their belief that the universe reveals itself in mathematical form, and that 'All is Number'.

The notion of putting the things of the world into some kind of ordered scheme is very ancient, and very powerful. Such ordered schemes or catalogues were central to early science, including the Pythagoreans. As the philosopher Jacob Klein explained in *Greek Mathematical Thought and the Origin of Algebra*, 'The general point of view governing the efforts of the Pythagoreans might be sketched out as follows: they saw the true grounds of the things in this world in their countableness, inasmuch as the condition of being a "world" is primarily determined by the presence of an "ordered arrangement" or *taxis*. [Conversely, any] order rests on the fact that the things ordered are delimited with respect to one another, and so become countable.'

The Pythagoreans' study of number was not entirely mathematical (at least as we understand that term), as they also proclaimed to find mystical meanings in numerical relations. Many religious and divinatory traditions associate particular qualities with particular numbers or patterns (e.g. astrology, tarot cards, or the *I Ching*), and these associations may feel far from arbitrary. For example, consider the following pairs of words:

Light/Dark; Warm/Cold; Right/Wrong;
Present/Absent; Active/Passive; On/Off;
True/False; Heavy/Light; Wet/Dry.

Now consider the following trinities:

Beginning/Middle/End; Viewer/Viewing/Viewed;
Past/Present/Future; Mind/Body/Soul;
Father/Son/Holy Ghost; Brahman/Shiva/Vishnu.

The mathematical fact of this matter is entirely mundane: counting the words in the first list proceeds 'one, two', while for the second list we have 'one, two, three'. But there is also a potent metaphorical connection within each group: the paired words have an oppositional character, with one part being defined in opposition to the other. In contrast, the trinities have a sense of interpenetration or interdependence. That is to say, something of the entire trinity is present in each part, and to separate the parts of each trinity is to divide an authentic whole.

As the integers were sacred to the Pythagoreans, they were motivated to grasp as much as they could of the character of each number, combining a mystical or metaphorical approach, together with what is still considered rational. Although it falls within a larger, extremely ancient tradition, Pythagorean number science was quite distinct from the knowledge that preceded it. A principle technique was to separate the integers into specific and intricately related categories, according to characteristics that can be demonstrated geometrically. For example, the Pythagoreans would speak about:

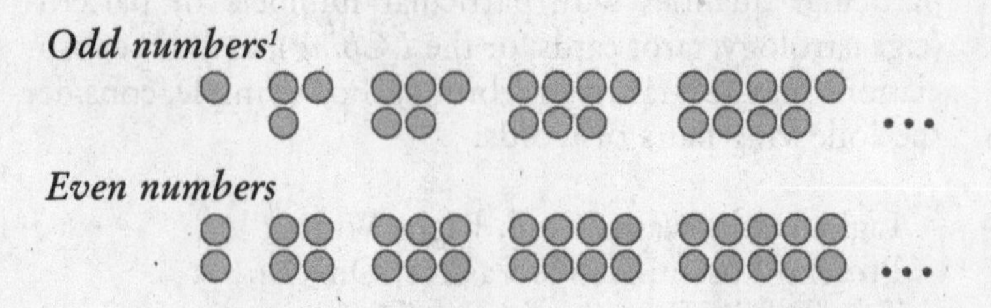

1 Modern mathematicians consider one to be an odd number, a square number, and so on, but the Ancient Greeks would not have included it in such lists. For them, numbers only began when there was a multitude of countable things.

Composite numbers
An integer is said to be composite if it is the multiple of two smaller integers (excluding one). Any composite number of dots can be arranged into a rectangle.

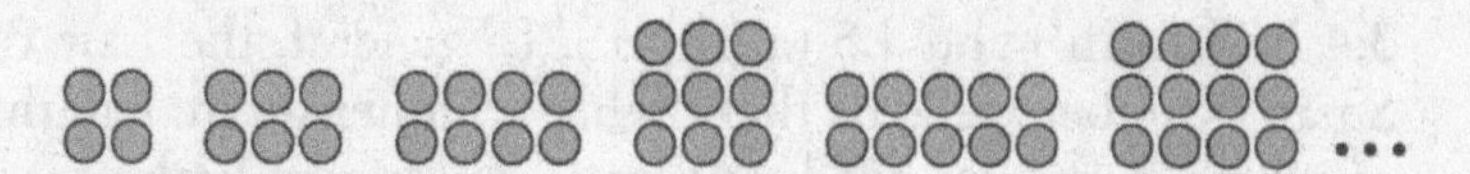

Prime numbers
By definition prime numbers are not composite, so they cannot be formed into any rectangle (other than a line).

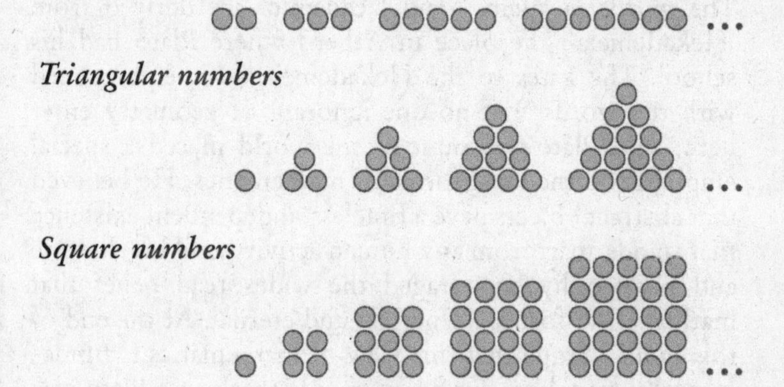

Triangular numbers

Square numbers

The Pythagoreans' belief in the supreme power of number found a beautiful confirmation in their study of stringed instruments. Halving the length of a plucked string raises the note by one octave, and people can clearly hear the harmony between two notes an octave apart. More generally, the pentatonic scale (black notes on a piano) can be produced by using the following sequence of lengths:

1 (doh), 8/9 (ray), 4/5 (me), 2/3 (soh), 3/5 (la), 1/2 (doh)

Similarly, the diatonic scale (white notes on a piano) runs as follows:

> 1 (doh), 8/9 (ray), 4/5 (me), 3/4 (fah), 2/3 (soh), 3/5 (la), 8/15 (te), 1/2 (doh)

These tunings are used because they generate a preponderance of the ratios 1:2 (an octave), 2:3 (a musical 'fifth'), 3:4 (a 'fourth') and 4:5 (a 'third'). In general, the Law of Small Numbers tells us that highly commensurate lengths of vibrating strings produce pleasing, resonant harmonies, while less commensurate lengths produce dissonance.

Plato and Symmetric Form

The words 'academy' and 'academic' are derived from 'Hekademeia': the place in Athens where Plato had his school. The gates to the Hekademeia were emblazoned with the words 'Let no one ignorant of geometry enter here,' and Plato's account of the world placed a special emphasis on the ideal forms of mathematics. He believed that abstract objects have a timeless, independent existence that stands apart from any human activity, and his rhetoric and philosophy encouraged the widespread belief that mathematical forms are perfect and eternal. At the end of this book I argue that this view of mathematics is fundamentally mistaken, but many people think that Plato was right, and describe themselves as Platonists or Realists.

The claim that mathematical forms are perfect and eternal has proved to be a potent and appealing notion. Indeed, there is a direct, historical link from Plato's rhetoric to the many spherical domes in churches, synagogues and mosques. More generally, it is striking that across the globe, from Mexico to Tibet, depictions of the cosmos have almost invariably been constructed along geometric lines. It would seem that when we conceive of a scheme, and relay that across generations, it tends to acquire an orderly, geometric form. For example, in many disparate cultures we find rings of hell; heavenly spheres; the four corners of the earth, and so on.

A fascination with symmetric, regular forms can be found in many cultures. What made the Ancient Greeks so remarkable was the systematic way that they investigated the properties of mathematical shapes, cataloguing and rigorously deducing their various properties. Long before Plato, people spoke of triangles, squares and other 'regular polygons', but as we shall see, our understanding of such forms made great advances in the time of Plato. By definition, a polygon is any planar figure with an integer number of straight sides. In a regular polygon, every side is the same length and every corner is identical. Now, a particularly ancient challenge is to find all of the different ways that we can cover a flat surface using regular polygons as tiles. In particular, imagine arranging a number of polygons so that they touch corners at one point.

If the polygons are going to completely cover a flat surface, the corners that meet at the given point must contain angles that add up to 360°. It follows that if we are to cover a flat surface with one type of regular polygon, the number of angles in one corner of that polygon must precisely divide 360°. Only three types of regular polygon satisfy this condition:

1. Triangular tiling. Six triangles can meet at a point because there is an angle of 60° between adjacent sides of an equilateral triangle, and 360° = 6 × 60°.
2. Square tiling. Four squares can meet at a point because there is an angle of 90° between adjacent sides of a square, and 360° = 4 × 90°.
3. Hexagonal tiling. Three hexagons can meet at a point because there is an angle of 120° between adjacent sides of a regular hexagon, and 360° = 3 × 120°.

The angle between adjacent sides of a regular pentagon is 108°, so where three pentagons meet we have a total of 324°, and a gap of 36°. Only two polygons with more than six sides can meet at a point, so triangles, squares and hexagons are the only regular polygons that can be used to tile the plane. If instead of using a palette of one kind of regular polygon we allow any combination of regular polygons while maintaining identical vertices (i.e. insisting that each corner is touched by an identical sequence of polygons), then there are exactly eight tiling possibilities.

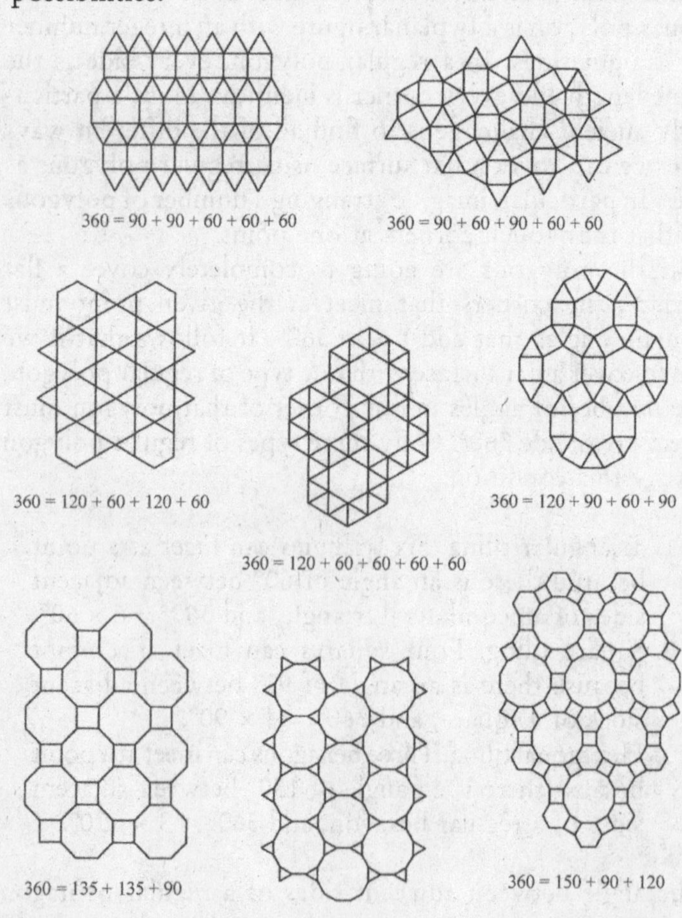

Plato studied mathematics with his friend and fellow Athenian Theaetetus, and it was Theaetetus who first identified all the different ways of enclosing a finite volume of space within a boundary composed from a single type of regular polygon. A cube is the most familiar example of a finite volume with a boundary composed of a single type of regular polygon, but – as we shall see – there are several others. Because they play a central role in Plato's metaphysics, shapes of this kind have somewhat unfairly acquired the name 'Platonic solids'. In order to find a complete list of these deeply regular shapes, Theaetetus needed to consider all the different types of corner that can possibly be constructed out of a single kind of regular polygon. That is to say, he needed to calculate the number and type of faces that could possibly meet at a point.

It should be clear that at each corner at least three polygons must meet. Two polygons or fewer cannot make a solid, volume containing form. Furthermore, when the corners of some polygons meet at a point, the total number of angles in those corners must be less than 360°. If the angles added up to 360°, the polygons in question would lie flat on the page, and make a tiling instead of a volumetric form. By falling short of 360° an arrangement of polygons can close in on a space, and may fit together to make one of the Platonic solids.

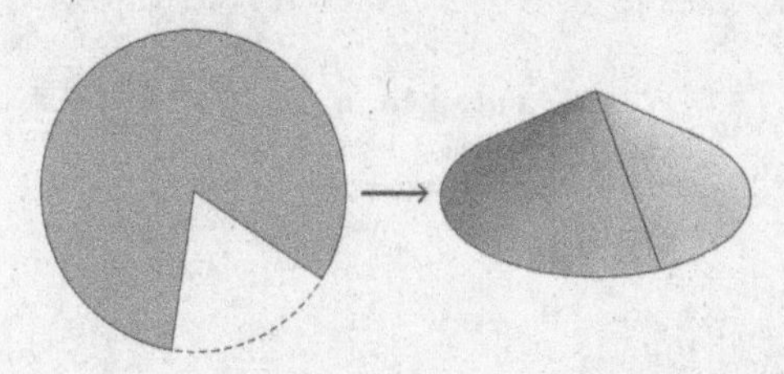

Given these constraints, there are only five possibilities for any corner of a Platonic solid. Three triangles can meet at each corner, and we can see that this is a genuine possibility because $3 \times 60° = 180°$, which is less than 360°. Similarly, four triangles can meet at each corner, and we can see that this is a possibility because $4 \times 60° = 240°$, which is less than 360°. Likewise, we can have five triangles meeting at each corner, but we cannot have six. If six equilateral triangles touch corners they lie flat on the page, which corresponds to the fact that $6 \times 60° = 360°$. To recap, if we use triangles to make the faces of a Platonic solid, we can have three triangles meeting at each corner, four triangles meeting at each corner, or five triangles meeting at each corner, but those are the only possibilities. Using square faces, the only possibility is that three of them meet at each corner (giving 270°), as four squares that touch at one corner lie flat on the page. We can also have three pentagons meeting at each corner, as that gives us a total of 324°, but that is the only remaining possibility. Hexagons and other polygons with more than five sides cannot produce a Platonic solid, as if you put three of these shapes together you get an angle that is at least as big as 360°.

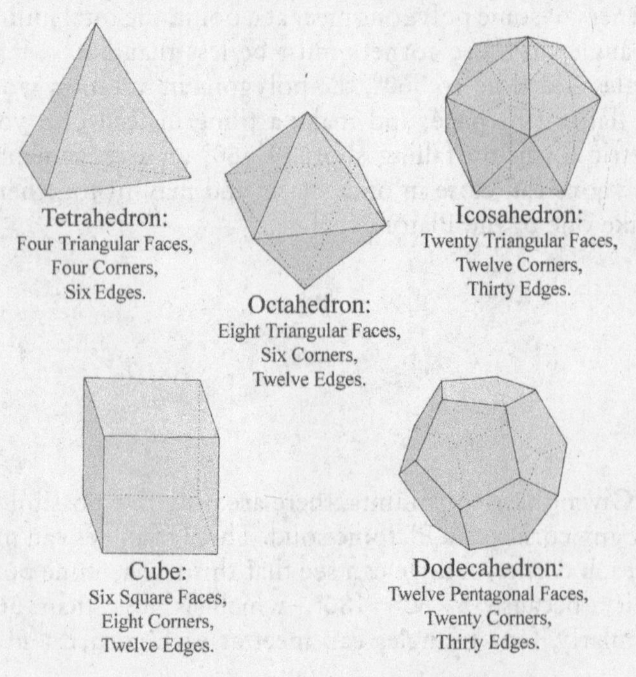

This argument does not complete the proof that there are precisely five Platonic solids. We have yet to prove that there are in fact shapes with vertices of the kinds described. Furthermore, we have yet to prove that there can only be one shape whose vertices satisfy the previous descriptions. For example, how do we know that there cannot be a shape that has three squares meeting at every corner, but which is different from a cube? Similarly, how do we know that there is a one-and-only shape that has three pentagons meeting at each of its corners? I will return to this point later in the book, but for the moment I think it will suffice to say that, greatly to their credit, the Greeks did not leave this final step as an assumption.

Euclidean Geometry

On 20 January 331 BC, Alexander the Great was sailing along the Egyptian coast, opposite the island of Pharos. Recognizing the many natural advantages of this site, he ordered that a city should be built in his name. Within decades, Alexandria was home to Macedonians, Greeks, Arabs, Babylonians, Assyrians, Italians, Carthaginians, Persians, Egyptians, Gauls, Iberians and Jews. To help establish the new metropolis as a great cultural centre, Alexander's successor and half-brother ordered that an enormous library be built. A great deal of money was spent, and scholars were employed to gather an encyclopaedic range of texts into an open, secular institution, modelled on the Athenian schools. It was in this newly founded city that Euclid (*c.* 325–265 BC) started his school of mathematics. Euclid himself trained at Plato's Academy, but under the rule of Alexander's general, Ptolemy I, Alexandria rose to become the scientific capitol of the world. Indeed, within Euclid's lifetime it became a greater centre of mathematical excellence than anywhere in Greece and, what is more, it maintained its pre-eminent position for five hundred years or so.

Euclid's *Elements* is the most influential textbook ever

written, and only the Bible has been printed in a greater number of editions. Although his book (or sequence of books) has been a major influence for more than twenty-three centuries, Euclid did not invent or discover the kind of geometry that bears his name. Nor did he invent the axiomatic method, whereby basic assumptions are listed and then deductions made from a definitive list of axioms through a logically structured sequence of theorems. The messy truth is that there was a long and complex evolution behind the emergence of Euclidean geometry, but Euclid's particular summary of the essential truths of Greek geometry is a *tour de force*, largely because of its exceptionally clear logical structure.

His books start with five famous axioms, and by employing these explicit statements we can logically clarify our language for describing shapes. Instead of relying on our intuitive interpretation of words like 'point', 'straight line', 'circle' or 'right angle', Euclid stated five defining principles for these things – the famous axioms of Euclidean geometry. In order to prove all of Euclid's theorems, we only need to refer to the following definitive properties of straight lines, points, right angles and so on.

1. There is precisely one shortest path between any two points. We refer to any such finite path as a straight line segment.
2. Any straight line segment can be extended indefinitely, forming a straight line.
3. Every straight line segment can be used to define a circle. One end of the segment is the centre of the circle, and its length forms the radius.
4. All right angles are essentially identical, in that any right angle can be rotated and moved to coincide with any other.
5. Given any straight line and a point that is not on that line, there is precisely one straight line that passes through the point, and does not intersect the line.

The fifth axiom effectively specifies what Euclid meant by the term 'parallel', and it also determines his interpretation of the term 'direction'. For millennia it was understood that two straight lines point in the same direction if and only if they do not cross (that is, if and only if they are parallel). In a later chapter I shall return to the subject of Euclid's fifth axiom, and elucidate the rethinking of geometry that took place in the nineteenth century. In particular, I shall discuss non-Euclidean geometry, and the significance of the fact that there are geometries that are fundamentally different to those described by Euclid.

First I want to point out that Greek geometry in general, and Euclid's *Elements* in particular, have exerted a profound influence: an influence that has reached well beyond the lives of mathematicians. Art, architecture, philosophy, theology, science and countless other endeavours have all been shaped by Euclid's example. It is therefore understandable that many philosophers, theologians and other constructors of arguments have echoed something of his form, thinking it only proper to start an argument with explicitly stated premises, and then use axioms to derive further implications.

As well as being a paradigm-defining example of a thoroughly sound argument, the practice of geometry also had an incalculable influence on the visual arts, particularly architecture. It is impossible to imagine how our world would have looked if it had been designed by people who did not use rulers and compass, as apart from a few notable exceptions, almost every blueprint, from medieval churches to modernist blocks, can be drawn using these two instruments. Furthermore, in classical geometry we often say things like, 'if we extend these lines we can see that they meet at a point'. Architects and painters have often favoured subtle kinds of order, where the generating pattern can be seen more clearly by extending the lines involved.

Straight lines and right angles are so prevalent in architecture that people who try using other shapes are often said to be 'experimenting with non-Euclidean geometry'. This is a perfectly understandable abuse of the term, but, strictly speaking, geometry is more fundamental and much harder to rethink than our palette of shapes or forms. Our geometry is present in the language that we use to describe our shapes, but the shapes themselves are not enough to determine which geometry we may be using. There is no problem with using the same geometry to study squares, circles or weird, irregular blobs. The geometry itself is characterized by such fundamental things as the kinds of symmetry we can possibly exhibit in the given space, or by the definitive relationships between straight lines, angles, distances and so on. In other words, a geometry is not defined by the shapes that we typically or most easily talk about. We can describe the same shapes using the language of Euclidean geometry or other essentially different, non-Euclidean geometries: the difference is not in the shapes, but in the meanings of the terms we use to describe them.

The Euclidean Algorithm

The concept of proportion is fundamental to art, architecture and mathematics. There is an ancient technique for characterizing or determining the ratio between given lengths, and although it predates Euclid this method is now known as the Euclidean algorithm. The first point to appreciate is that proportion is a more abstract concept than a ratio of lengths, a ratio of areas, and so on. For example, if we have a piece of string that is 20 cm long and another that is 40 cm long, we have a ratio of 20 cm:40 cm, but we also find it obvious that one is twice as long as the other. If we consider the distances 17 miles and 34 miles, then although the actual lengths are completely different, we recognize that 20 cm:40 cm is in the same *proportion* as 17 miles:34 miles. In both cases we have the proportion 1:2, and we understand that the expression 1:2 stands as a representative for every equally proportioned ratio. Also note that in writing a proportion we do not need units, so the proportion 1:2 could equally well refer to a ratio of lengths, areas, or any other quantified property.

In his famous book, Euclid described the well-known method for determining the proportion between two given lengths. The first step is to construct a rectangle whose sides possess the lengths we are interested in. Having done that, we can determine the ratio between the lengths of those two sides by employing the following technique:

1. Draw the largest square that fits inside your rectangle.
2. Squeeze in as many of these squares as you possibly can. If you can fill the entire rectangle with your squares, you are finished. Otherwise, there will be a rectangular space that has not been covered by squares. We now take the remaining rectangle and apply step 1.

The Euclidean algorithm terminates when we find a square that fits perfectly, filling in the entire rectangle. Squares of this size can be used to make a grid that perfectly subdivides the generating shape, and this will be the coarsest square grid that can possibly be used to divide the rectangle into squares. In other words, the width of the smallest square is equal to the greatest common divisor of the lengths of the two sides.

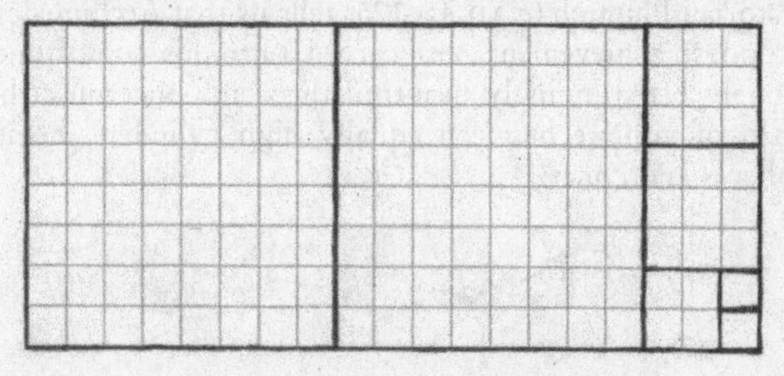

For some rectangles, there simply isn't any square grid that will fit perfectly. In this case our algorithm can be continued indefinitely. If the lengths of the sides are integers with no factors in common (e.g. 5 cm x 9 cm), the final step of our algorithm uses 1 cm squares. If the numbers have a common factor (e.g. 3 cm x 9 cm), we produce a scaled up version of some co-prime pattern (e.g. we finish with 3 cm squares):

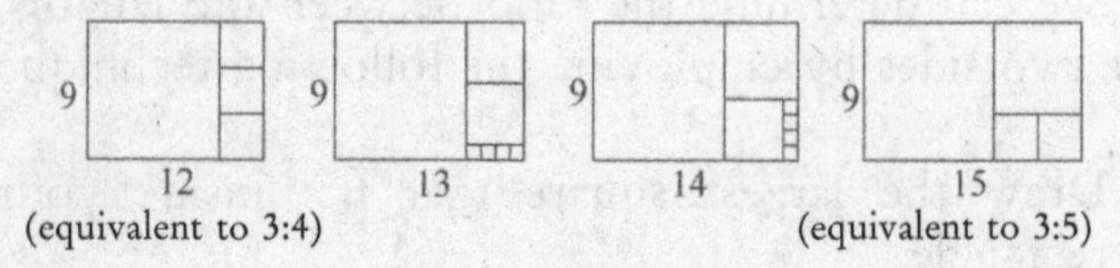

This technique makes the proportions of the generating rectangle evident and intelligible, and it has inspired countless generations of mathematicians, artists and architects.

Archimedes

Archimedes of Syracuse (*c.* 287–212 BC) is arguably the greatest mathematician to ever live. He has also been described as the first mathematical physicist, as he developed theories that accurately predict when a thing will balance, and when a thing will float. As well as his astounding theoretical accomplishments, Archimedes was a brilliant engineer: people have been using the Archimedean screw to move water uphill for over twenty-two centuries, and it is still used to this day. However, the Ancient Greek historian Plutarch (*c.* AD 45–125) tells us that Archimedes' proudest achievement was carved onto his tombstone, namely his stunningly beautiful theorem concerning the ratio of volume between equally high cylinders, hemispheres and cones:

Archimedes became famous in his own lifetime, and we know that at least one contemporary account was written about his life and work (now sadly lost). Plutarch, Livy, Cicero and Vitruvius all mentioned his life's work, and he was helped in acquiring legendary status by defending the Sicilian city of Syracuse. Plutarch writes of the fear and respect that Archimedes struck in the Roman invader's hearts, as he fought them from the city for over two years: 'If they only saw a rope or a piece of wood extending beyond the city walls, they took flight, exclaiming that Archimedes had once again invented a new machine for their destruction.'

His conceptual grasp of the physical literally changed the world, and shaped the way that technology developed. For example, when he wrote *On the Equilibrium of Planes*, Archimedes gave a completely axiomatic treatment of mechanics, enabling logical deductions for physical systems. By such means he devised a powerfully general explanation of the functioning of rollers, wedges, levers and pulleys, using terms whose relevance clearly extends beyond the particular contraptions that are physically present. His strictly mathematical approach enabled the calculation of centres of gravity, and he was in a position to make such general statements as 'The tipping point is where the centre of gravity is directly above the edge of the base.' Similarly, he was the first person to calculate when two objects will balance on a see-saw.

As if that wasn't enough to ensure his immortal fame, Archimedes was the first person to prove that a sphere of radius r must have a volume of $4/3\pi r^3$, and a surface area four times greater than a circle of equal radius (i.e. $4\pi r^2$). He proved many fundamental results, but a particularly attractive example of his work can be found in the way that he considered the area of a circle. By definition of the term 'π', the circumference of a circle of radius r is equal to $2\pi r$.

Archimedes knew that if we slice a circle into equally sized segments and rearrange those component shapes, the total area will not change. He also knew that given any circle, we can draw a hexagon that is just small enough to fit inside the circle, and we can draw another hexagon that is just large enough to contain the entire circle. The area of our circle must be bigger than the area of the smaller hexagon, and smaller than the area of the larger hexagon. Furthermore, it is easy to calculate the area of these hexagons because they are composed of triangles of known size. Archimedes understood that in principle we can do the same for a polygon of any number of sides, and as we consider polygons with larger and larger numbers of sides, we find shapes whose areas are closer and closer to the area of a circle. This method can be used to calculate estimates of π. Indeed, Archimedes managed to calculate π to two decimal places by using a 96-sided polygon. More importantly, Archimedes' ingenious arguments showed that the area of circle of radius r must be πr^2. Like all mathematical theorems, this fundamental truth has many different proofs, including the following beauty devised by Leonardo da Vinci:

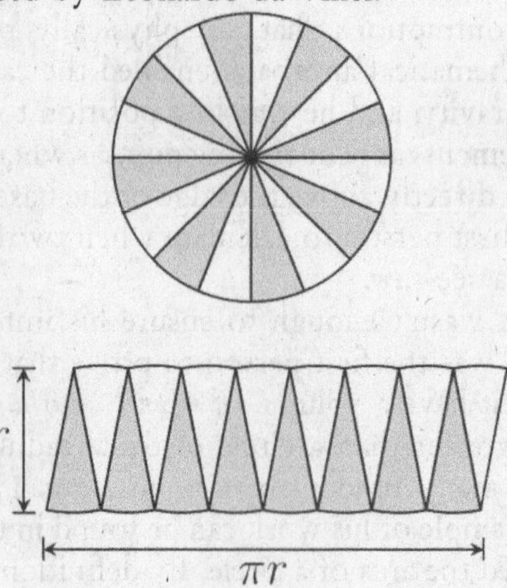

If we slice a circle of radius r into an increasing numbers of segments, those segments can be arranged into a shape that is increasingly similar to a rectangle of width πr and height r.

In many of his proofs, Archimedes argued that a given area or volume cannot be bigger than some given number, and likewise it cannot be smaller than the given number. For example, Archimedes proved that the area of a circle cannot be bigger than πr^2, and it cannot be smaller than πr^2. It follows that the area of a circle must be equal to πr^2: a form of argument known as 'proof by exhaustion'. The mathematician and astronomer Eudoxus (*c.* 408–355 BC) is credited with being the first person to prove that one quantity is equal to another by demonstrating that it cannot be bigger or smaller, and over the following millennia, proof by exhaustion showed itself to be a neat but rather unproductive idea. However, as we shall see in Chapter 5, the thought of covering a shape with increasingly fine pieces is an ancient ancestor of the most fruitful mathematical idea of the modern era: the infinitesimal calculus.

Alexandria in the Age of Rome

In the centuries that followed the founding of Alexandria, Rome grew to dominate all of its surroundings. As Rome's sphere of influence expanded, countless manuscripts were 'donated' or pillaged to furnish the libraries of eminent Romans. For example, we know that Cleopatra gave Caesar a gift of tens or hundreds of thousands of manuscripts from the Royal Library of Alexandria. Fortunately, many Romans revered the mathematicians of Greece, appreciating both the practical and theoretical significance of mathematical knowledge. There was a popular trade in busts and portraits of the famous mathematicians, and Greeks were often employed as tutors. The Roman Empire spread the practice of mathematical education throughout Europe (well beyond the ethnically Roman), but surprisingly the cosmopolitan city of Alexandria remained dominant as the centre for purely mathematical study.

Despite the unfortunate fate of the tragically flammable Royal Library, the city of Alexandria remained a magnet for mathematicians for many, many centuries. We have good reason to believe that Archimedes studied there, and it was in Alexandria that the Pythagorean scholar Nicomachus of Geresa (*c.* AD 60–120) wrote his *Introduction to Arithmetic*, employing arithmetic notation and ordinary language to explain topics that Euclid had described from an essentially geometric perspective. This book was highly influential, and remained the standard European textbook on arithmetic for more than a thousand years. Some fifty years after *Introduction to Arithmetic* was published, Claudius Ptolemy (*c.* AD 85–165) produced a work that was even more influential: *The Almagest*, or the 'Great Collection'.

As a Roman citizen of Egypt working in Alexandria, Ptolemy employed the detailed astronomical knowledge of the Babylonians to make improvements on Eudoxus's model of planetary motion. The resulting masterpiece was considered to be a definitive guide to the movement of the heavens, and over the following millennia, countless people used his system to predict successfully the apparent location of the planets in the night sky. *The Almagest* also spread the system of longitude and latitude, and it contained many sophisticated arguments about the relationship between measurements of angles and measurements of length (what we call trigonometry).

Following the collapse of the Roman Empire, priests and laymen continued to teach and study the mathematics of Greco-Roman civilization. The diplomat and philosopher Anicius Manlius Severinus Boethius (*c.* AD 480–525) deserves particular mention in this regard. His *Institutio Arithmetica* was not a great piece of original mathematics, but his reverence and love for the subject helped to ensure that the Catholic Church would teach and preserve the Pythagorean knowledge of number, following Boethius' educational *quadrivium* of arithmetic, geometry, astronomy and music.

There is a certain poetry in the fact that while so many practical points of reference were lost (e.g. the means of making concrete), the most abstract of notions remained unbroken. Indeed, a long line of Christian theologians, including Saint Augustine, have supported the view that God's creation is mathematical. It is therefore understandable that despite the terrible upheavals following Rome's collapse, crucial fragments of mathematical science were faithfully preserved. By the time of the Middle Ages, scholastic writers were making tentative advances from the body of established mathematical knowledge. However, until Italian mathematicians started to make exciting advances in the late fifteenth century, ancient works such as Euclid's *Elements* were still considered to be the pinnacle of mathematical knowledge.

In later chapters we will see how Europe experienced a renaissance in mathematics, building on the achievements of Greek and Arab scholars. First, I want to focus on an aspect of reality that has occupied mathematicians, artists and architects for millennia. The notions of measurement and proportion are exceptionally ancient, and in the next chapter I will explain how counting and continuous measurement have been used to make proportion evident. In particular, we will see how the insights of Eudoxus and Dedekind extended the concept of number, from integers to fractions and on to irrational numbers.

Chapter 3: RATIO AND PROPORTION

> 'The concept of number is the obvious distinction between the beast and man. Thanks to number, the cry becomes a song, noise acquires rhythm, the spring is transformed into a dance, force becomes dynamic, and outlines figures.'
>
> Joseph de Maistre, 1753–1821

Measurement and Counting

In daily life, we don't just count objects, we also measure quantities such as length, area, weight and time. First we select a unit of measurement: feet, acres, grams or hours as the case may be. We assign the relevant unit quantity the measure of one, and then count the number of unit measures that make up the quantity to be measured. For example, we might measure the length of a field by *counting* the number of feet between one end and the other. In general, the process of counting out units may not 'come out even'. For example, our field might be longer than sixty feet, but shorter than sixty-one feet. In that case we may measure the remaining distance by using some subunit measure, obtained by subdividing our original unit into n equal parts.

The Ancient Egyptians and numerous other cultures

studied this kind of process. In ordinary language the standard units and subunits are given their own names, as feet are subdivided into inches, hours are subdivided into minutes, and so on. In general we divide our unit measure into n subunits, and when we measure something we need to count out m of these subunits to make up the quantity to be measured. In that case we say that we have measured a fraction m/n. Note that the 'denominator' n tells us what kind of subunit we are using, while the 'numerator' m tells us how many of the subunits we have counted.

Over the course of several centuries, the ancient equivalents to the modern symbols m/n gradually lost their association with the process and units of measurement. In other words, people began to consider fractions as 'pure' numbers, much like the integers. How do we justify extending the word 'number' from the counting numbers onto the fractions? Well, we might say that just as it doesn't matter whether you are counting apples, pears or people, it doesn't matter whether we are measuring fractional quantities of distance, weight, time, etc. It is also absolutely fundamental that we can add and multiply the fractions alongside the counting numbers. Indeed, the rules for adding and multiplying fractions can be summarized as follows:

$$\frac{a}{b}+\frac{c}{d}=\frac{ad+bc}{bd}, \quad \frac{a}{b}\times\frac{c}{d}=\frac{ac}{bd}$$

$$\frac{a}{a}=1 \text{ and } \frac{a}{b}=\frac{c}{d} \text{ if and only if } ad=bc.$$

From a modern perspective we accept the fractions as a legitimate number system because the rules for adding and multiplying fractions satisfy the same, axiomatic laws as the addition and multiplication of integers. More specifically, the following statements are true whether p, q and r are integers or fractions:

$$p + q = q + p, p + (q + r) = (p + q) + r, pq = qp,$$

$$p(qr) = (pq)r \text{ and } p(q + r) = pq + pr.$$

For many centuries it was believed that the only conceivable quantities were fractional or 'rational'. However, as I shall show in the next section, this plausible claim is not in fact true. As an initial observation, consider the following diagrams:

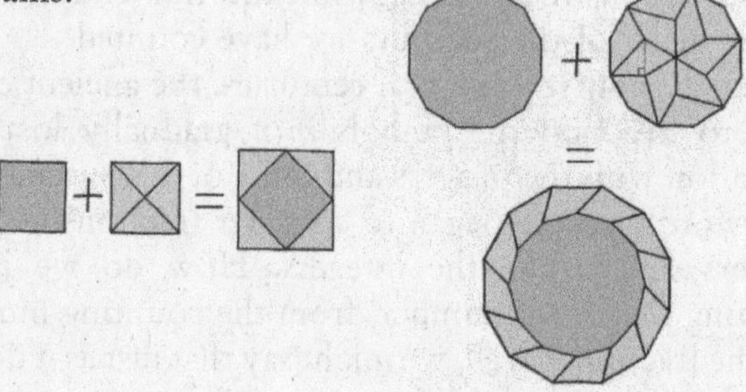

In each case we have combined a pair of identical shapes to produce a similar shape with double the area. The sides of these larger shapes are $\sqrt{2}$ times greater than the sides of the original, component shapes. We can prove this fact using Pythagoras' Theorem. More generally, we can use Euclid's axioms to prove that rescaling a two-dimensional shape by making the lengths n times longer changes the shape's area by a factor of n^2. Because the second, larger shapes have double the area ($n^2 = 2$), the lengths of the sides must have increased by a factor of $\sqrt{2}$. Similarly, we can combine three hexagons to produce one with triple the area, and sides $\sqrt{3}$ times that of the originals:

Because the ratios 1:√2 and 1:√3 are geometrically constructible, we intuitively feel that √2 and √3 must be numbers of some kind. In other words, we readily accept that each of these things has a single, definitive size. This notion is extremely ancient, and hundreds of generations from across the globe have pondered the question, 'How long is the diagonal of a square, given the length of the sides?' Indeed, a Babylonian clay tablet from *c.* 1800–1600 BC tells us that this ratio of lengths can be expressed as:

$$1 \text{ to } 1 + \frac{24}{60} + \frac{51}{60^2} + \frac{10}{60^3}$$

(a solution that is exact enough for any practical purpose, as the error in this approximation is only 0.00004%).

As we saw in the previous chapter, Pythagorean science used the language of the integers, and in that language we might say that a quantity simply *is* a ratio of two integers. However, during the fifth century BC, the Ancient Greeks proved that √2 cannot be expressed as a ratio of integers, and nor can √3 or √*p* where *p* is any prime.

Reductio Ad Absurdum

In order to prove that √2 cannot equal any fraction, we must first prove the Fundamental Theorem of Arithmetic. Both of these proofs can be established by the mathematician's gambit: *reductio ad absurdum*. As the mathematician G. H. Hardy remarked, 'It is a far finer gambit than any chess gambit: a chess player may offer the sacrifice of a pawn or even a piece, but a mathematician offers *the game*.' I shall return to this idea, but for now let us just accept that *we do not say contradictory things about mathematical objects*. In short, we cannot work with a logically inconsistent system, so if a putative set of axioms is inconsistent but we want to continue doing mathematics, we have to change one of our assumptions. Somewhat surprisingly, this negative principle (the ban on inconsistency) enables us to prove positive new conclusions, including facts concerning the relationship between fractions and measured lengths.

Euclid's proof of the Fundamental Theorem of Arithmetic begins with an axiomatic observation: 'Every collection of positive integers contains a smallest number.' In particular, suppose that there are integers that are neither prime nor the multiple of primes (integers with property P, say). It follows from our axiom that if there are any integers with property P, there must be a smallest integer (called k) that has the property P. In other words, if the Fundamental Theorem of Arithmetic were false, there would have to be an integer k, which is the smallest number that is neither prime nor the multiple of primes. If k is not the multiple of any two integers (apart from k x 1 of course), then by definition k is prime, which would mean that it does not have property P after all.

The other possibility is that there are two integers s and t, such that s x $t = k$. In that case, k must be larger than either s or t. This implies that s and t cannot have property P, which means they must be prime or the multiple of primes. This follows because by definition k is the smallest number with property P. But the multiple of any two multiples of primes is itself a multiple of primes, which means that k cannot have property P. Since there cannot be an integer k with property P, it must be true to say that every integer can be factorized.

Reductio ad absurdum is a simple trick, but it reveals a wealth of truth: every integer must either be prime or it must be equal to the product of some primes. In other words, given any positive integer N (other than $N = 1$), there is a collection of prime numbers $p_1, p_2, \ldots, p_n$ such that $N = p_1, \times p_2 \times \ldots \times, p_n$. Euclid also proved that for every integer N, there can only be one collection of primes that multiply together to give the number N. For example, $2 \times 3 = 6$ and $3 \times 2 = 6$, but every other multiplication of primes equals something other than 6.

Using a similar method, we can prove that $\sqrt{2}$ is not equal to any ratio of integers. Suppose for the sake of argument that we have found a fraction N/M, which satisfies the equation $N/M = \sqrt{2}$. We can rewrite this equation by multiplying both sides by M and then squaring both sides, giving us the equation $N^2 = 2M^2$. By the Fundamental Theorem of Arithmetic, we must be able to rewrite the integers N and M as a sequence of prime numbers. We therefore have something of the form:

$$(p_1 \times p_2 \times \ldots \times p_n)^2 = 2(q_2 \times q_2 \times \ldots \times q_m)^2, \text{ or equivalently}$$

$$p_1^2 \times p_2^2 \times \ldots \times p_n^2 = 2 \times q_1^2 \times q_2^2 \times \ldots \times q_m^2.$$

If it is indeed the case that the left-hand side equals the right, then the prime factors of the left-hand side must be identical to the prime factors of the right. But it cannot be true that both sides have the same prime factors, because the factors of the left-hand side are two lots of p_1, two lots of p_2, and so on up to two lots of *pn*, while on the right we have two lots of q_1, two lots of q_2, and so on up to two lots of *qm*, plus a single factor of 2. This solitary factor of 2 cannot have a counterpart on the left-hand side, which means that the two sides of the equation cannot be equal to one another. We must therefore reject our assumption that for some pair of integers N and M we have $N^2 = 2M^2$. In other words, we must conclude that there cannot be a fraction $N/M = \sqrt{2}$.

The proof that $\sqrt{R}$ cannot equal any fraction when R is any prime is very similar. First we suppose that we have found a fraction N/M, which satisfies the equation $N/M = \sqrt{R}$, which is equivalent to supposing $N^2 = RM^2$. The next step is to rewrite N and M as a sequence of prime numbers, giving us something of the form:

$$(p_1 \times p_2 \times \ldots \times p_n)^2 = R(q_2 \times q_2 \times \ldots \times q_m)^2, \text{ or equivalently}$$

$$p_1^2 \times p_2^2 \times \ldots \times p_n^2 = R \times q_1^2 \times q_2^2 \times \ldots \times q_m^2.$$

As in the case where $R = 2$, the solitary factor of R on the right-hand side cannot have a counterpart on the left-hand side, which means that there cannot be a pair of integers N and M such that $N^2 = RM^2$. Hence, there cannot be a fraction $N/M = \sqrt{R}$.

Once the Greeks had constructed this beautiful little argument, they were faced with a considerable challenge. How could they maintain the link between proportion (measuring lengths against one another) and the familiar laws of arithmetic (counting, adding and multiplying)? Given that $\sqrt{2}$ is irrational (i.e. given that $\sqrt{2}$ cannot be expressed as a ratio of integers), how can we justify calling it a number? How can we justify the assumption that we can add and multiply $\sqrt{2}$ together with the fractions? To put it another way, how can we numerically express a geometric length when we know that the length in question does not equal any fraction?

Eudoxus, Dedekind and the Birth of Analysis

Thanks to the widespread use of graphs most people find the concept of a number line highly intuitive. The basic idea is that every point on a number line represents a number: a subtle form of metaphor with a long and complex history. Although graphs as we know them are a relatively recent innovation, the mathematician and astronomer Eudoxus (*c.* 408–355 BC) understood that although a continuous number line necessarily contains more than just fractions, it is made intelligible through a rational framework. Eudoxus's insight was one of the most pivotal events in the history of mathematics, and to understand the mystery from which this clarity emerged, let us return to Pythagoras' Theorem. We have already seen that whenever we have a right-angled triangle whose sides are of length a, b and c, it must be the case that $a^2 + b^2 = c^2$. Conversely, *every* solution to the equation $a^2+b^2=c^2$ corresponds to a right-angled triangle with sides of length a, b and c.

We can be certain that every solution to this equation corresponds to an 'actual' triangle because we can draw two lines at right angles, and let the lengths of those lines equal any pair of positive numbers a and b. When we connect our endpoints, the hypotenuse that we draw must have a length equal to our third number c, because we have already specified that our triangle is right angled, and so by Pythagoras' Theorem the length of the hypotenuse must satisfy the relationship $a^2+b^2=c^2$.

People could draw and identify right-angled triangles long before the time of Pythagoras. Indeed, some people believe that as far back as 1800 BC the Babylonians knew that if a triangle was formed by joining lines of length a, b and c, then the triangle will be right angled whenever $a^2+b^2=c^2$. The main evidence for this claim is a clay tablet known as Plimpton 322, which systematically lists fifteen pairs of integers $\{a, c\}$. In every case there is some integer b, such that $a^2+b^2=c^2$. For example, the tablet contains the pairs of integer $\{45, 75\}$, $\{1679, 2929\}$ and $\{12709, 18541\}$, which form 'Pythagorean triples' with the integers 60, 2400 and 13500.

The connection between the geometric facts of right-angled triangles and purely arithmetic facts such as $3^2+4^2=5^2$ has greatly impressed hundreds of generations. On first inspection there appears to be little connection between arithmetic and geometric knowledge, for as John Stillwell writes in his excellent *Mathematics and its History*:

> Arithmetic is based on counting, the epitome of a *discrete* process. The facts of arithmetic can be clearly understood as outcomes of certain counting processes, and one does not expect them to have any meaning beyond this. Geometry, on the other hand, involves *continuous* rather than discrete objects, such as lines,

curves, and surfaces. Continuous objects cannot be built from simple elements by discrete processes, and one expects to see geometric facts rather than arrive at them through calculation.

Despite the many differences between arithmetic and geometry, Pythagoras' Theorem hints at a depth of interconnection. Indeed, his now legendary contribution was to demonstrate that the following statements are *logically equivalent*:

1. A triangle whose sides are of length a, b and c (where c is the longest length) is in fact a right-angled triangle if and only if
2. A square whose sides are of length c contains the same area as a square of side a plus a square of side b.

We could say and understand these individual sentences long before we could prove their logical equivalence, so Pythagoras' achievement is a very striking one. It should also be noted that as a *formula* the theorem reads as a statement about the addition and multiplication of quantities, because the lengths of the given sides are numbers that are related by the formula $a^2 + b^2 = c^2$. The Greeks certainly knew how to add and multiply fractions, but *the lengths mentioned in the formula may well be irrational*, as one can certainly consider triangles with irrational sides. For this reason (and many other related ones), it was imperative that the process of adding and multiplying arbitrary quantities was understood, and related back to the standard methods for adding and multiplying fractions.

This problem is far less daunting to the modern reader, as most people think of a number as being something like 2.713... (a possibly infinite string of digits). With such a representation, one can naturally extend the operations of addition, subtraction, multiplication and division in a relatively straightforward way. For example, suppose that we want to find $x = 2.713\ldots \times 3.425\ldots$ It isn't very difficult to work out how to calculate each of the digits of x in turn: you just need to be systematic about it. Indeed, just by looking at the first digits we can deduce that x is bigger than 2×3 and smaller than 3×4. By looking at the second digits we can work out that x is bigger than 2.7×3.4 and smaller than 2.8×3.5, and so on. Crucially, if you want to find a finite number of digits of x, you only need to look at finitely many digits to the right of the equals sign. My point is that when we imagine doing sums with an infinite string of digits, we simply keep on adding and multiplying in the time-honoured way, but accept that when we are using infinite decimals, there is quite literally no computational end in sight, even for a single multiplication.

That is all well and good, but the central problem remains. Is it possible to prove that two infinite decimals are equal (as they may be in Pythagoras' Theorem)? To make sense of this question, we must first be clear about the defining relationships greater than, smaller than and equal to. For fractions, we have a very simple criterion for equality:

$$\frac{a}{b} = \frac{c}{d} \text{ if and only if } ad = bc.$$

For example, we say that $1/2 = 3/6$ precisely because $1 \times 6 = 2 \times 3$. Similarly, we say that:

$$\frac{a}{b} < \frac{c}{d} \text{ if and only if } ad < bc.$$

In other words, we can compare rational numbers by multiplying and comparing integers. Eudoxus realized that not only are there well-defined criteria for the comparison of fractions, there is also a reasoned sense of measure that applies to every 'real number'. The crucial and deceptively simple observation is that two real numbers (points on a continuous number line) are said to be 'different' if and only if there is a gap between them. If there is a gap, then there must be a fraction that is bigger than one of the numbers, and smaller than the other, because every stretch of the number line contains fractions. This means that two real numbers are in fact one and the same if and only if it is impossible to find a fraction that is larger than one number and smaller than the other.

It is highly significant that equality between real numbers is defined in terms of computational failure (that is, the failure to find a fraction that is bigger than one number and smaller than the other). On a more positive note, we can show that two different definitions pick out the same real number by demonstrating that being larger than one definition and smaller than the other is a logical impossibility for any *fraction*. In other words, we can prove equality between real numbers within a rational framework, and we don't need to worry about irrational numbers messing things up.

The other crucial point about Eudoxus's brilliant idea is that in order to speak of a number's size (that is, its defining characteristic), all that we require is the ability to say if it is bigger, smaller or equal to any given fraction. For example, we can compare any fraction f to $\sqrt{2}$ by calculating f^2 and comparing this number to 2. Our ability to add, multiply and compare fractions is firmly rooted in our ability to count, as the proper way to add, multiply and compare fractions is strictly determined by the proper way to add, multiply and compare integers. We can therefore be absolutely certain that every fraction rightly belongs in one of two sets: the set of 'smaller than $\sqrt{2}$' fractions, and the set of fractions 'at least as big as $\sqrt{2}$' (which is exactly the same as the set of fractions bigger than $\sqrt{2}$).

The idea of partitioning the fractions into 'smaller' and

'larger' is critically important. Among other things, a rule for sorting any fraction into either 'smaller than $\sqrt{2}$' or 'larger than $\sqrt{2}$' provides us with a strictly deterministic method for generating the decimal expansion of $\sqrt{2}$. The procedure runs something like this:

> $1^2 = 1$ which is less than 2, so 1 must be less than $\sqrt{2}$.
>
> $2^2 = 4$ which is greater than 2, so 2 must be greater than $\sqrt{2}$.
>
> $1.4^2 = 1.96$ which is less than 2, so 1.4 must be less than $\sqrt{2}$.
>
> $1.5^2 = 2.25$ which is greater than 2, so 1.5 must be greater than $\sqrt{2}$.
>
> $1.41^2 = 1.9881$ which is less than 2, so 1.41 must be less than $\sqrt{2}$.
>
> $1.42^2 = 2.0164$ which is greater than 2, so 1.42 must be greater than $\sqrt{2}$, …

There are many other techniques that we can use to effectively sort any fraction into 'smaller' or 'larger'. For example, consider the following method for calculating whether some fraction a/b is more or less than π, bearing in mind that π is normally defined as the circumference of any circle divided by the diameter of that circle, but π is also equal to the area of a circle with radius one.

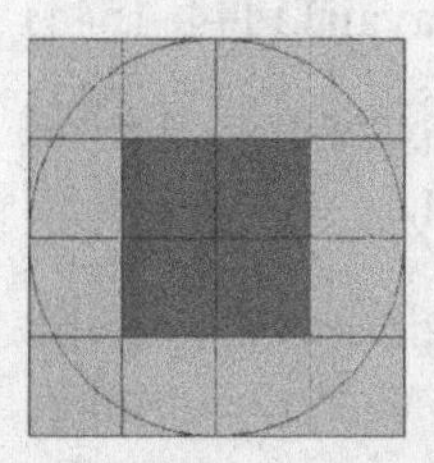
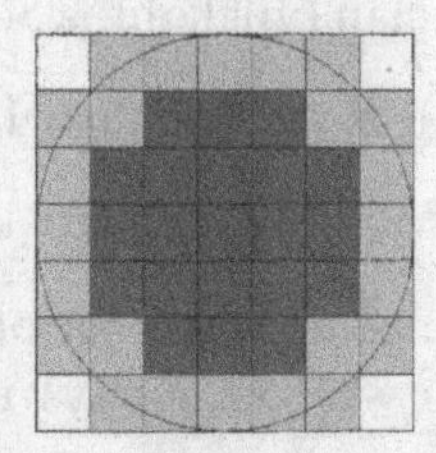
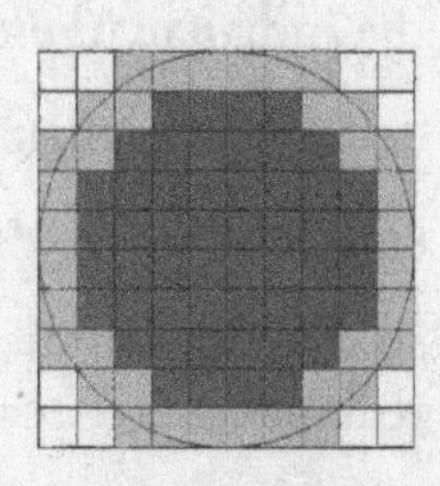

We can calculate π by drawing circles on grids of increasingly tiny squares. These squares are coloured white if they lie completely outside the circle, black if they lie inside the circle, and grey if the circle crosses them. As we use finer and finer grids, the black area gets closer and closer to π, while the grey area becomes arbitrarily small.

If a fraction a/b is different to π (as every fraction is), then one of these pixellated images will be detailed enough to show this fact. As Eudoxus understood, 'a/b is different to π' simply means 'there is a detectable gap between a/b and π'. That is to say, the sufficiently fine images will demonstrate that one (and only one) of the following statements is true:

1. The area of the black region is larger than a/b, so a/b is smaller than π, or
2. The area of the black region plus the area of the grey region is less than a/b, so a/b is larger than π.

Notice that in each case we are comparing our fraction a/b against other fractions (the relevant areas). We can correctly fill in any grid by using Pythagoras' Theorem, which tells us the distances between the corners in the square grid and the centre. Hence our procedure for checking the relative size of a/b and π depends on nothing more or less than the knowledge of how to add and multiply integers.

Another way to specify a particular real number is to write it as the 'limit case' of an infinite sum. For example, the Indian mathematician Nilakantha Somayaji (1444–1544) demonstrated that $\pi/4$ is smaller than 1, bigger than $1 - 1/3$, smaller than $1 - 1/3 + 1/5$, bigger than $1 - 1/3 + 1/5 - 1/7$, and so on. This sequence of approximations defines a specific real number because we can always use a finite number of these approximations to tell whether any given fraction is greater or smaller than π. It should also be noted that many different sequences can specify precisely the same real number, as the following equations indicate:

Nilakantha Somayaji (*c.* 1500):

$$\pi = 1 - \frac{1}{3} + \frac{1}{5} - \frac{1}{7} + \frac{1}{9} - \frac{1}{11} + \frac{1}{13} - \frac{1}{15} + \ldots$$

François Viète (1593):

$$\frac{2}{\pi} = \sqrt{\frac{1}{2}} \times \sqrt{\frac{1}{2} + \frac{1}{2}\sqrt{\frac{1}{2}}} \times \sqrt{\frac{1}{2} + \frac{1}{2}\sqrt{\frac{1}{2} + \frac{1}{2}\sqrt{\frac{1}{2}}}} \times \ldots$$

John Wallis (1655):

$$\frac{\pi}{2} = \frac{2}{1} \times \frac{2}{3} \times \frac{4}{3} \times \frac{4}{5} \times \frac{6}{5} \times \frac{6}{7} \times \frac{8}{7} \times \ldots$$

Leonhard Euler (*c.* 1750):

$$\frac{\pi^2}{6} = \frac{1}{1^2} + \frac{1}{2^2} + \frac{1}{3^2} + \frac{1}{4^2} + \frac{1}{5^2} + \frac{1}{6^2} + \frac{1}{7^2} + \ldots$$

$$\frac{\pi^2}{6} = \frac{2^2}{(2^2 - 1)} \times \frac{3^2}{(3^2 - 1)} \times \frac{4^2}{(4^2 - 1)} \times \frac{5^2}{(5^2 - 1)} \times \frac{6^2}{(6^2 - 1)} \times \ldots$$

Recurring Decimals and Dedekind Cuts

When Eudoxus clarified the fundamental relationships 'greater than', 'smaller than' and 'equal to', mathematics was essentially geometric in character. However, later generations have taken a more algebraic approach, and by the nineteenth century mathematicians were once again wanting to clarify precisely what is meant by a 'real number'. Intuitively speaking, real numbers correspond to points on a number line. As we shall see, this is a rather subtle concept, despite its intuitive appeal. The modern definition of the real number line is due to Richard Dedekind (1831–1916), but the idea that numbers could be identified as points on a line was first articulated by René Descartes (1596–1650).

We know exactly how to do arithmetic with some of the points on a number line (e.g. the integers), but the problem with saying that every point on a number line is actually a number is that it isn't obvious how we are supposed to do arithmetic using an arbitrary point on a geometric line. Dedekind recognized that there is a better way to define what is meant by a real number, which did not appeal to the geometric notion that a line is the path traced by a moving point. Dedekind insightfully observed that every real number can be identified with a pair of *sets*, namely the set of fractions smaller than x, and its inverted twin, the set of fractions at least as big as x. This complementary pair of sets is called a 'Dedekind Cut', for obvious reasons.

It may be more intuitive to think of real numbers as being points on a number line, but when rigour is called for mathematicians clarify the notion of 'a point on a number line' by turning to the formalism of Dedekind Cuts. That is to say, specifying a real number x is understood as being the same thing as specifying a way to divide the set of fractions into two, such that every fraction in the 'smaller' set is smaller than every fraction in the 'larger' set. By definition a number x is bigger than y if and only if there is a fraction that is in the set 'larger than y', which is also in the set 'smaller than x'. Similarly, the number x is equal to y if and only if the set 'smaller than x' is identical to the set 'smaller than y'.

Now, when we first learn about fractions we tend to think of one number being divided by another (fractions as a kind of verb). By gaining experience of the addition and multiplication of fractions, we learn to think of them as nouns ('completed things' in their own right). We must make a similar conceptual transition from verb to noun when we specify a number using recurring decimals (e.g. 0.3333...). The idea behind this notation is that we can use an infinite sequence of terms (0.3, 0.33, 0.333, ...) to pick out one particular number. That is to say, we can specify one particular Dedekind Cut by putting fractions in the 'smaller' set if and only if they are smaller than one of the terms 0.3, 0.33, 0.333, ... This leads us to our next question: are there any numbers that have regular, recurring digits, but which do not equal any fraction? In particular, could $\sqrt{2}$ have recurring digits, or must it continue forever un-repeating?

We can answer this question by appreciating some of the algebraic qualities of recurring decimals. Consider, for example, the number 0.234523452345... By naming this number x, we can make the following argument:

$1000x = 2345.23452345...$ and
$x = 0.23452345...$ therefore
$9999x = 2345$ (subtract the second equation from the first).

Dividing both sides by 9999 gives us the equation

$$x = 0.2345... = \frac{2345}{9999}.$$

We can shift the recurring digits along by adding an appropriate number of zeroes to the denominator. For example, $0.00234523435 \ldots \frac{2345}{999900}$. We can also find an equivalent fraction when there is a finite string of digits before the recurring ones. For example,

$$0.6623452345 \ldots = \frac{66}{100} + \frac{2345}{999900} = \frac{662279}{999900}.$$

It should now be clear that every recurring decimal can be written as a fraction. This means that $\sqrt{2}$ definitely cannot have recurring digits, because unlike numbers with recurring digits, $\sqrt{2}$ cannot be written as a fraction. One particular application of this argument highlights the intuition refining power of Eudoxus's conception of equality. When $x = 0.999...$ we have:

$$10x = 9.9999... \text{ and}$$
$$x = 0.9999... \text{ therefore}$$
$$9x = 9 \text{ and so } x = 1.$$

Many people are somewhat perturbed to find that the recurring decimal 0.999... is equal to one. But we can see that this is true, because if a fraction is smaller than one of the numbers 0.9, 0.99, ... , then it must also be smaller than one. Furthermore, the number one is the smallest fraction that is not smaller than any of the fractions 0.9, 0.99, 0.999, ... This implies that one is the smallest member of the set 'larger than 0.999...' Hence the symbols 1 and 0.999... indicate precisely the same Dedekind Cut, which is another way of saying that they are one and the same real number.

The notion that there is an infinitesimal difference between 0.999... and 1 is not a part of ordinary mathematics (though there are forms of mathematics that include infinitely small quantities). In other words, the number one is the only real number that 0.999... can equal. It may seem strange to say that the difference between 1 and 0.999... is actually zero, but in the words of the great mathematician Leonhard Euler, 'To those who ask what the infinitely small quantity in mathematics is, we answer that it is actually zero. Hence there are not so many mysteries hidden in this concept as they are usually believed to be.' More generally, our analysis shows that every recurring decimal is equal to some fraction, and by considering the process of long division we can see that the converse of this statement is also true. In other words, every fraction has an infinitely repetitive decimal expansion, and every recurring decimal is equal to some fraction.

Continued Fractions

Decimal notation is an extremely powerful and convenient way to represent numbers. However, it is far from being the only system, and there is an alternative form of representation, which is particularly elegant and interesting. The forms I am referring to are known as continued fractions, and many historians believe that they were studied by the Ancient Greeks (who were certainly familiar with the key ideas). Although he didn't claim to invent them, our earliest unambiguous record of the use of continued fractions comes from a book called *L'Algebra*, which was written by Raphael Bombelli in 1572.

So what are these things called continued fractions? Well, we are all familiar with the idea that one divided by two is a number, which can be written in the form $\frac{1}{2}$. We are also familiar with the idea that if $\frac{1}{2}$ is a number, then so is $2+\frac{1}{2}$. But since $2+\frac{1}{2}$ is a number, does it not follow that one divided by this number is also a number, namely $\frac{1}{2+\frac{1}{2}}$? And why shouldn't we continue adding integers and dividing the resulting number into one? For example, it should be clear that $\frac{1}{3+\frac{1}{2+\frac{1}{2}}}$ is also a specific number, it has just been written in an unusual manner.

Numbers of this form are called continued fractions. In the case of a finite chain we can always rewrite our number as an ordinary fraction, but that may not be the case if our numbers form an infinite chain. As we are most used to thinking of arbitrary numbers in terms of an infinite sequence of decimals, it is instructive to think about how we can convert a decimal number into the form of a continued fraction. The fundamental observation is that every real number x can be written in the form $\lfloor x \rfloor + \Delta(x)$ where $\lfloor x \rfloor$ is the integer part of x and $\Delta(x)$ is some remaining real number between 0 and 1. For example, the integer part of 2.269 is 2, while the remainder $\Delta(2.269)$ is 0.269. A greater insight comes from observing that for every non-integer real number x, there is an integer part $\lfloor x \rfloor$ but *there is also an integer part corresponding to the real number* $\frac{1}{\Delta(x)}$.

For example, when $x = 2.269$ we have an integer part 2. The remainder is 0.269, and $\frac{1}{0.269} = 3.71$ ... The integer part of 3.71 ... is 3, and this tells us that the number 2.269 is close to $2\frac{1}{3}$. By an iterative process of inverting a sequence of remainders and taking integer parts, we can generate a specific sequence of integers that represent our real number x. In this example the first integer is 2 and the second integer is 3, but more generally for every real number x there is an integer $\lfloor x \rfloor$ and a (possible infinite) sequence of positive integers r_1, r_2, ... such that:

$$x = \lfloor x \rfloor + \cfrac{1}{r_1 + \cfrac{1}{r_2 + \cfrac{1}{r_3 + \dots}}}$$

Writing a number x in terms of the above notation is known as 'expressing x as a continued fraction'. Furthermore, finding this sequence of integer parts is equivalent to applying the Euclidean algorithm. In other words, given a real number x, we can find the corresponding continued fraction by drawing a 1:x rectangle, and then systematically filling in as many squares as we can, using the largest possible square at each step.

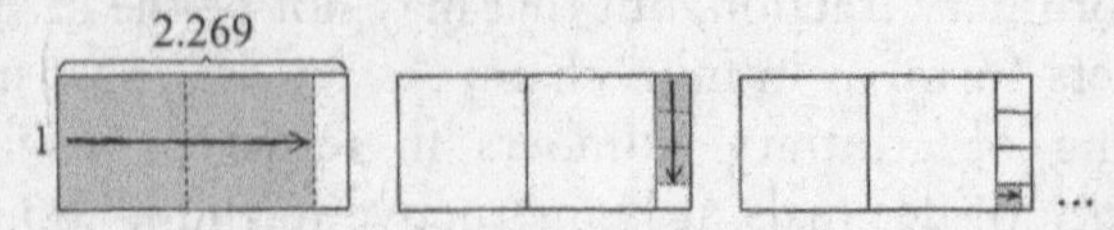

When $x = 2.269$ we have $\lfloor x \rfloor = 2$, and we can fit two squares inside the rectangle.

We can fit three squares into the space that remains, which tells us that $r_1=3$.

We can only fit one square in the space that remains, which tells us that $r_2=1$.

If there is a length that exactly divides both sides of the rectangle, then our procedure comes to a halt when we find that length. This happens if and only if the number x is expressible as the ratio of two integers. As a representation of number continued fractions are rather like decimal fractions, in the sense that the first numbers tell us the most about the overall size of x. Furthermore, the rational sequence $2, 2+\frac{1}{3}, 2+\cfrac{1}{3+\frac{1}{1}}$ is comprised of increasingly accurate approximations for our real number x = 2.269, just as the sequence 2, 2.2, 2.26, ... is composed of increasingly accurate approximations for the underlying number 2.269 (say).

We can rewrite the terms $\lfloor x\rfloor, \lfloor x\rfloor+\frac{1}{r_1}, \lfloor x\rfloor+\cfrac{1}{r_1+\frac{1}{r_2}}$ as ordinary fractions, and as the sequence progresses the terms become arbitrarily close to x. This is crucial, because it means that sequences such as these determine one particular Dedekind Cut. Furthermore, the odd-numbered terms in the sequence are all at least as small as x. For example, our first estimate $\lfloor x\rfloor$ must be at least as small as x. Similarly, the rounding down in the third, fifth and seventh approximations produces a number that is at least as small as x. Conversely, all the even-numbered terms are at least as big as x. This follows because $\lfloor x\rfloor$ is at least as small as x, which means that $\frac{1}{\lfloor x\rfloor}$ must be at least as big as $\frac{1}{x}$.

Quadratic Equations and the Golden Ratio

Finite continued fractions can be rewritten as ordinary fractions, which means that finite continued fractions can be used to specify precisely the same real numbers as recurring decimals. But what kind of real number can be specified by an infinitely recurring continued fraction? Consider the following example:

$$1+\cfrac{1}{1+\cfrac{1}{1+\cfrac{1}{1+\dots}}}$$

This is a highly exalted number, and in keeping with tradition I shall label it ϕ (or phi). By definition, the number ϕ satisfies the equation $\phi = 1+\frac{1}{\phi}$. If we multiply both sides by ϕ_2, we get $\phi^2 = \phi + 1$. We can solve this equation using the standard 'quadratic formula', which tells us that $\phi = \frac{1+\sqrt{5}}{2}$, or 1.618033989 … Alternatively, the equation $\phi = 1+\frac{1}{\phi}$ can be used to generate its own solution (as a continued fraction), by an iterative process of repeated substitution. We simply replace each occurrence of ϕ with $1+\frac{1}{\phi}$, and by doing this over and over again we generate the continued fraction

$$\phi = 1+\cfrac{1}{1+\cfrac{1}{1+\cfrac{1}{1+\dots}}}$$

The ratio 1:1.618… (or equivalently 0.618… : 1) is called the 'golden ratio', and rather like π it seems to crop up everywhere. Many natural patterns exhibit the golden ratio, as do many of the world's most famous works of art. For example, the Ancient Greeks believed that the ideal human form should be full of golden ratios (e.g. the ratio from navel-to-toe to head-to-toe 'should' be golden), the Mona Lisa was carefully proportioned by the repeated use of the golden ratio, and the modernist Le Corbusier advocated the use of the golden ratio when designing buildings or furniture.

I will return to this particular example of a continued fraction presently. First I want to point out a very general connection between quadratic equations and continued fractions with a repetitive structure. For example, consider the number

$$x = 2 + \cfrac{1}{1 + \cfrac{1}{2 + \cfrac{1}{1 + \dots}}}$$

Because this continued fraction has a repetitive structure, we can replace the second occurrence of $2 + \frac{1}{1 + \dots}$ with an x, to give us the equation

$$x = 2 + \frac{1}{1 + x}.$$

It is possible to prove that given any equation that describes a repetitive, continued fraction, we can algebraically rearrange that equation into the form $ax^2 + bx + c = 0$, where a, b and c are ordinary fractions. Equations of this form are known as 'rational quadratic equations'. In our particular example of a repetitive continued fraction, some elementary algebra can prove that $x^2 - 2x - 2 = 0$, so we have $x = 1 + \sqrt{3}$. In other words:

$$\sqrt{3} = 1 + \cfrac{1}{1 + \cfrac{1}{2 + \cfrac{1}{1 + \cfrac{1}{2 + \dots}}}}$$

We have just seen that repetitive continued fractions can solve rational, quadratic equations. When he was only seventeen, the wildly original and tragically short-lived Evariste Galois (1811–1832) demonstrated that there is also a converse to this argument. In other words, it is an algebraic fact that a real number x can be the solution to a rational quadratic equation if and only if x can be written in the form of a repetitive continued fraction. For example, suppose that we start with the equation $x^2 = 2$. We are free to add an x to both sides of the equation, because this operation preserves equality. Furthermore, $x + x^2 = 2 + x$ is equivalent to $x(1 + x) = (1 + x) + 1$. Dividing both sides by $(1 + x)$ gives us $x = 1 + \frac{1}{1 + x}$. By repeatedly substituting $1 + \frac{1}{1 + x}$ for every occurrence of x, we effectively home in on the only positive solution to the equation $x^2 = 2$, which is $1 + \cfrac{1}{2 + \cfrac{1}{2 + \dots}}$ or 1.41421.

Galois' other great achievement was even more remarkable. The night before he was killed in a duel, he spent his time writing an explanation of what is now known as 'Galois Theory'. His conclusion was that equations that involve a term x^5 are fundamentally different to equations that only involve x, x^2, x^3 or x^4. As you may recall, if we are given any equation of the form $ax^2 + bx + c = 0$, we can use a simple formula to find all the values of x that solve our equation (namely $x = \frac{-b \pm \sqrt{b^2 - 4ac}}{2a}$). Similarly, there is an equation for finding the solutions to cubic equations, and there is also an equation for finding the solutions to equations involving x^4. Many mathematicians had tried to find a general rule for solving equations in x^5, but Galois is famous for proving that an equivalent formula cannot possibly exist.

Structures of Irrationality

If a real number is expressible as the ratio of two integers, we say that it is a rational number. Every real number is either rational or irrational, but some numbers are more irrational than others. That is to say, some irrational numbers are even less like fractions than other irrational numbers. To understand why this is so, imagine picking a point on a number line, then trying to find a rational approximation that is both simple and accurate. For every integer n, we can find the best (or joint best) estimate of the form m/n.

$1/n$ $2/n$ x $3/n$ $4/n$

It should be clear that every real number is at most $\frac{1}{2n}$ away from one of these fractions, and this maximum error occurs when x is halfway between $\frac{m}{n}$ and $\frac{m+1}{n}$. Also notice that if x is close to halfway between $\frac{m}{n}$ and $\frac{m+1}{n}$, that means it is close to $\frac{m+\frac{1}{2}}{n} = \frac{2m+1}{2n}$.

If we use larger and larger values of n, we can find increasingly accurate approximations to our number x. Furthermore, for every integer K there is a 'best rational approximation' of the form m/n, where $m \leq K$. The previous statement must be true because there are finitely many 'sensible' approximations of this form, so there must be at least one approximation that is as close to x as any of them.

For any particular integer K there will be some numbers that are very close to their best rational approximation, while other numbers will be relatively distant from their best rational approximation (where we insist that any approximation m/n is such that $m \leq K$). Later in this chapter we will see that some numbers are poorly approximated for every integer K. In the case of these highly irrational numbers, every integer ratio is a relatively poor approximation for the actual number.

To understand the structure of irrationality a little better, we need to appreciate a particular fact about the sequence of approximations generated by the Euclidean algorithm. Suppose that we generate the approximation m/n by our method of taking integer parts. It is possible to prove that for any fraction a/b, one of the following must be true:

1. m/n is closer to x than a/b, or
2. a is greater than m.

In other words, given a real number x, the Euclidean algorithm generates approximations of x that are the best rational approximations. The next step is to appreciate the pivotal relationship between the accuracy of these approximations, and the size of the next integer in the continued fraction. For example, $r_1 = \left\lfloor \frac{1}{\Delta(x)} \right\rfloor$ is large whenever $\Delta(x)$ is small. Furthermore, if the remainder $\Delta(x)$ is small, our initial approximation $\lfloor x \rfloor$ must have been close to x. More generally, the larger the value of r_n, the smaller the difference between r_{n-1} and $r_{n-1}+\frac{1}{r_n}$, and the smaller the difference between consecutive approximations.

As we work our way along the sequence of approximations generated by the Euclidean algorithm, we find an under-estimate, followed by an over-estimate, followed by an under-estimate, and so on. Because we alternate between under- and over-estimates, the difference between the n'th approximation and the true value of x must be less than the difference between the n'th approximation and the n+1'th approximation. It follows that if the integer is large, the n'th remainder must be small, and the n'th approximation must be very accurate.

As an example, $3+\frac{1}{7}=\frac{22}{7}$ is a pretty good approximation of π. We might say that π is well approximated by single-digit fractions, because most real numbers don't have such a simple but accurate approximation. Correspondingly, the third fraction approximation of π is $3+\cfrac{1}{7+\cfrac{1}{15}}$. One fifteenth is fairly small, so improving our estimate by including this factor results in a relatively small decrease in the overall size of our approximation. This implies that our original estimate must have been close to π (as indeed it was). On the other hand, π is poorly approximated by fractions with two-digit denominators, in the sense that most numbers can be approximated more accurately. This corresponds to the fact that the fourth digit in π's continued fraction is a one (as are almost half of the first million integers of π's continued fraction).

In a far more extreme case, it occurred to the Indian mathematician Srinivasa Ramanujan (1887–1920) that the number $e^{\pi\sqrt{163}}$ must be almost exactly an integer. In fact, there are seventeen digits before the decimal point, followed by thirteen zeroes. Correspondingly, the second term in the continued fraction form of $e^{\pi\sqrt{163}}$ is a whopping 1,333,462,407,511.

The Fibonacci Sequence

Leonardo Fibonacci (*c.* 1170–1250), also known as Fibonacci of Pisa, was an Italian merchant who travelled extensively, studying with Arab scholars. His influential book *Liber Abaci* was the first European textbook to cover decimal notation, and the now familiar techniques of multiplication and long division (operations that are much, much easier to carry out with decimals than Roman numerals). Fibonacci also helped to spread the basic ideas of algebra: a development that we shall examine in the next chapter. Ironically, most people have heard of Fibonacci because in 1877 a number theorist called Edouard Lucas was studying the sequence 1, 1, 2, 3, 5, 8, 13, … , and he decided to pay tribute to Fibonacci by calling that sequence the Fibonacci sequence. Fibonacci himself said very little about this sequence, but he would have recognized that it was the solution to one of the more whimsical problems he included in his book. That is to say, Fibonacci once posed the following question: If we start with a single pair of baby rabbits, how many pairs of rabbits will we have each month assuming that:

1. There are no deaths, and
2. Every month each pair produces one new pair, which becomes productive in the second month after birth.

The answer for each successive month can be found in the following sequence:

At each stage we add the two preceding terms to get the next in the sequence. The previous term accounts for all the rabbits that are surviving, and the term before that is equal to the number of mature pairs. As each mature pair produces a new born pair each month, the total of these two terms is equal to the total number of pairs of rabbits. As we move along this sequence, the ratio between successive terms gets closer and closer to 1.618... (the golden ratio). This is related to the fact that terminating the continued fraction $\phi = 1 + \cfrac{1}{1 + \cfrac{1}{1 + \dots}}$ after finitely many steps generates the following sequence of best rational approximations for ϕ:

$$\frac{1}{1}, \frac{2}{1}, \frac{3}{2}, \frac{5}{3}, \frac{8}{5}, \frac{13}{8}, \frac{21}{13}, \frac{34}{21}, \dots$$

Both the numerators and the denominators form the sequence we found with Fibonacci's rabbit problem, and so the sequence of integers 1, 1, 2, 3, 5, 8, 13, 21, ... is known as 'the Fibonacci sequence'. Returning to the golden ratio, the above sequence of rational approximations to ϕ have a fascinating and highly significant property. Because 1/1 is the largest fraction of the form $1/n$, the infinite part of our continued fraction (the bit we effectively ignore by stopping after finitely many steps) makes as big a difference as possible. In other words, every single step changes our approximation by the maximum possible amount, given our method of taking integer parts. This tells us that we have the maximum possible gap between every consecutive pair of under- and over-estimates, which means that none of our approximations are very good.

Because our method of generating continued fractions gives us the best possible approximations (for their size), the golden ratio really is the hardest number to approximate using fractions. This means that ϕ is the most irrational number of all! To reveal the beauty of this truth, consider a kind of automated drawing machine. These machines operate in a step-by-step fashion, moving a single shrinkable arm that pivots around the centre of each drawing. At each step the arm rotates through some fixed angle R, shrinks by some fixed proportion P, and then leaves a mark. To make the point clearer, I have also drawn lines around the edge of a containing circle to show the different directions that the arm was pointing in as it left the marks.

If we use a fraction or an integer for our angle R the marks lie in straight lines. Indeed, when $R = \frac{1}{n}360°$ we generate n radial lines. More generally, when $R = \frac{m}{n}360°$ (and m and n have no factors in common), we generate n radial lines.

Using rational values of R generates straight lines, while using an irrational angle guarantees that no two marks will lie on a straight line through the centre. This follows because finding two marks that lie on a straight line through the centre is equivalent to finding three integers a, b and c, such that:

$$(a \times R) = (b \times R) + (c \times 180°).$$

But in that case we would have $R = \frac{c}{a-b}180°$, which can only be true if R is rational. Although there is this fundamental difference between the patterns produced by rational and irrational values of R, similar values produce similar patterns, so every finite drawing looks like a (more or less) twisty version of some fraction. Reasonably straight lines indicate the presence of a 'good' fraction approximation:

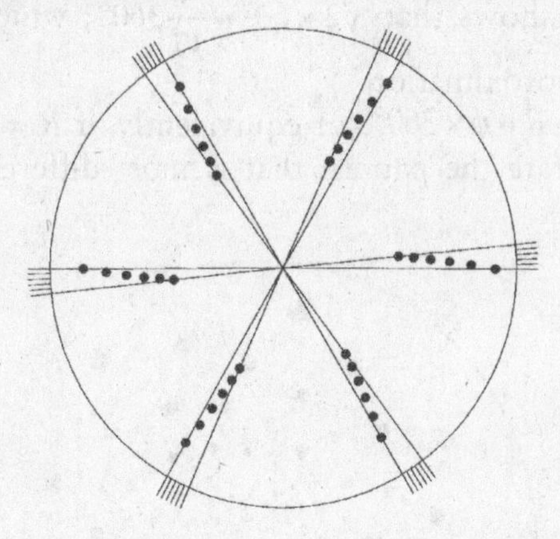

$R = 60.2°$, just over one sixth of a circle.

If the length of the arm stays relatively constant (i.e. if P is relatively close to one), there will be more points near the edge of our containing circle, so the eye tends to pick out a larger number of spiralling lines. This means that if we increase P we tend to notice that R is similar to some new fraction (one with a larger denominator). If R is very similar to the fraction in question, then the spiralling lines are relatively straight, and the marks around the edge of our containing circle are closely bunched.

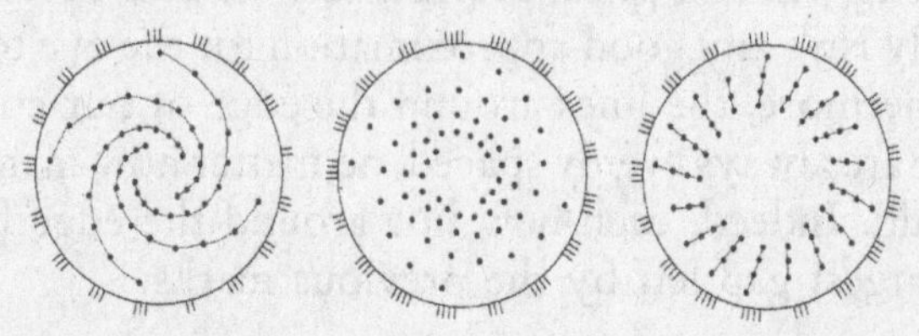

$R = \sqrt{2} \times 60°$ for all three drawings, and P increases from left to right.

The left most drawing has four spiral arms, illustrating the approximation $\sqrt{2} \times 60° = 84.85 \approx \frac{1}{4} 360°$. The third drawing shows that $\sqrt{2} \times 60° \approx \frac{4}{17} 360°$, which is a much better approximation.

When $R = \phi \times 360°$ (or equivalently, if R = 137.508 ...), we generate the pattern that is most different from any fraction:

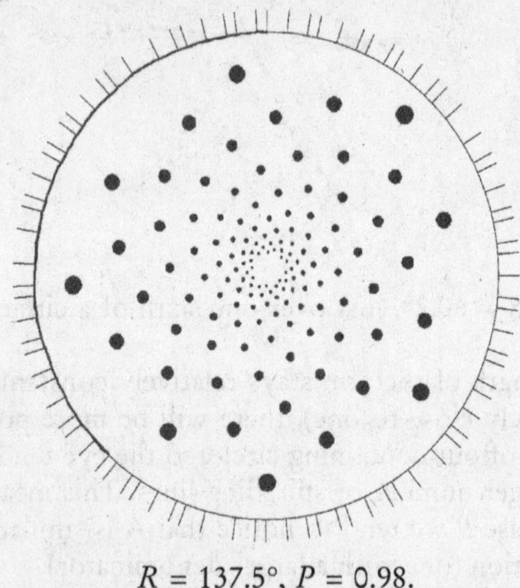

R = 137.5°, P = 0.98.

When $R = \phi \times 360°$ we can use any value for P and our drawing will still spiral dramatically. That is because there simply isn't any good approximation for the eye to notice! Furthermore, the lines around the edge of our containing circle are always evenly spaced, no matter how many points we add. Indeed, each new line around the edge falls into the largest gap left by the previous marks.

Many plants avoid growing branches, petals or leaves directly above one another, and such arrangements frequently contain golden ratios. Sunflowers, pineapples, raspberries, daisies and pinecones all exhibit particularly clear patterns that are based on the golden ratio. If you count the number of clockwise spiral lines formed by the pieces of these things, you (almost) always find a number in the Fibonacci sequence. The number of anticlockwise spiral lines will be an adjacent number in the sequence. Because the total number of pieces is the same whichever direction you count, the average number of points along the clockwise spirals will be in a golden ratio to the average number of points along the anticlockwise spirals.

People have marvelled at the natural preponderance of golden ratios for millennia. In particular, since the early nineteenth century biologists and mathematicians have been intrigued by the phenomena of spiral phyllotaxis: the golden arrangement of leaves, flowers or other lateral organs sticking out of a central axis or stem. It is clearly inefficient to grow a leaf directly above another leaf because one would put the other in the shade, so we might expect that plants would evolve non-overlapping leaf arrangements. Furthermore, the non-overlapping patterns that plants form ought to be ones that can be extended to accommodate more and more leaves, as plants grow more leaves over time.

As we have seen, growing each leaf at a golden angle relative to the one below is an excellent strategy, as it ensures that no matter how many leaves you grow, none of them will be directly above the leaves you have already grown. However, it is far from obvious that selective pressure to avoid overlapping necessarily produces the golden ratio, rather than one of the many other non-overlapping arrangements. It is only in recent years that the key mechanisms determining the location of leaves have begun to be understood, and, in particular, we now know that the plant hormone auxin plays a critical role. Cells with high levels of auxin tend to grow more rapidly, and high levels of auxin tend to induce the production of a leaf. Furthermore, cells actively pump the auxin they contain though special export channels, and the export channels of each cell tend to face whichever neighbouring cell contains the most auxin.

As a result of this process cells that contain relatively high levels of auxin tend to accumulate even more, while the neighbouring cells tend to deplete their stores. Furthermore, when a cell acquires higher and higher levels of auxin, a leaf begins to form. This process effectively drains the auxin from the neighbouring cells, inhibiting the growth of other leaves in the immediate vicinity. As the stem continues to grow, it is unlikely to grow a leaf directly above another leaf, because that part of the stem will have grown out of a region where the auxin concentration is particularly low (since the cells in that region are busy pumping auxin back towards the newly forming leaf). The fact that some cells grow faster than others means that there are asymmetric mechanical forces as the growing cells push against their neighbours, and in many plants the interplay between growth dynamics and auxin transport means that the places where leaves will grow tend to be related to one another by the golden angle.

Of course, nothing in biology is mathematically exact, but the patterns produced by many different plants closely resemble the patterns of the golden angle. Indeed, through a combination of computer models, physical models and biological experiments, it is becoming increasingly clear that the relationship between the golden angle and non-overlapping works both ways. Not only does the golden angle produce patterns with minimal overlaps, it is also the case that growth procedures that avoid overlapping have a strong tendency to produce golden angles.

We will have more to say about the role of mathematical models in biological science. But before we can appreciate the importance of this exciting new branch of knowledge, we should return to the beginnings of mathematical science. In particular, we should examine two ideas that have fundamentally shaped the way we think about the world: decimal notation and algebraic equations. Over the course of the next chapter I shall discuss the invention of zero, and we will see how the key ideas of algebraic number theory passed from India to the Muslim world, before reaching Europe.

Chapter 4:
THE RISE OF ALGEBRA

'There are the nine figures of the Indians: 9 8 7 6 5 4 3 2 1. With these nine figures, and with the sign 0 which in Arabic is called zephirum, any number can be written, as will be demonstrated.'

First lines of the *Liber Abaci*, written by Leonardo Fibonacci in 1202

Zero and the Position System

The modern system of writing numbers as a sequence of digits is so simple and convenient that we tend to take it for granted, but it is difficult to overstate the importance of this brilliant innovation. As the mathematician Alfred North Whitehead (1861–1947) once wrote:

> Before the introduction of [decimal notation], multiplication was difficult, and the division even of the integers called into play the highest mathematical faculties. Probably nothing in the modern world could have more astonished a Greek mathematician than to learn that, under the influence of compulsory education, the whole population of Western Europe, from the highest to the lowest, could perform the operation of division for the largest numbers. This

> fact would have seemed to him a sheer impossibility. ... Our modern power of easy reckoning with decimal fractions is the most miraculous result of a perfect notation.

Crucially, you cannot have decimal notation without a symbol for zero. The idea that nothing can be viewed as a number is a rather strange and subtle concept, but it is terribly important. Indeed, it is no exaggeration to say that mathematics, and therefore science, technology and culture, could not possibly be what they are today without the number zero. The number zero as we understand it today originated in India, but historians now believe that the proper place to start the story of zero is in Ancient Mesopotamia, where a succession of civilizations used clay tablets to keep numerical records. A wide range of number systems have been used over the ages, but a major advance occurred around the third dynasty of Ur (*c.* 2100 BC), when the Sumerians began to use position as part of their numerical system.

The nameless scribe who first began to use positional numeration was a real genius, and this exceptionally convenient idea is central to the modern decimal system. For example, we all know that '23' represents two lots of ten plus three, while '32' represents three lots of ten plus two. This demonstrates the fact that our notation is positional, which means that changing the relative position of our digits yields a different number. The same digit '3' can stand for three (as in '23'), thirty (as in '32'), three hundred (as in '302'), and so on. This contrasts with Roman numerals or the notation used by the Ancient Greeks, where the symbol V always stands for five, the symbol X always stands for ten, and so on.

There is, however, one crucial difference between modern numeration and the earliest position based systems: they didn't have a symbol like '0'. To understand how this exceptionally important change occurred, it is instructive to consider something like the counting boards or *abaci* that almost every civilization has developed. That way we can appreciate how people could do without such a symbol, and how helpful it was when the symbol was first introduced. The Sumerians used a system based around the number sixty, rather than base ten. Indeed, the fact that we divide circles into 360 degrees, hours into 60 minutes and minutes into 60 seconds is part of our inheritance from Mesopotamian civilizations. However, for the sake of simplicity, let us consider the example provided by a Madagascan method for counting armies.

What would happen is that soldiers were led in single file, and as each one passed, a pebble was dropped into a bowl. Once the bowl contained ten pebbles it was emptied, and a single pebble would be dropped into a second bowl, used for counting groups of ten soldiers. Similarly, once the second bowl contained ten pebbles it was emptied, and a single pebble would be dropped into a third bowl, used for counting groups of a hundred soldiers. Now, imagine that the authorities wanted to keep a record of how the counting process had gone. This record might have read 'The third bowl held six pebbles, the second bowl was empty, and the first bowl held four pebbles.' This account would tell the reader that there were six hundred and four troops. Note that whatever words or notation were used to say that the second bowl was empty played an essential role, and this role is now performed by the symbol '0'. However, counting and record keeping in this manner does not require us to think of zero as a proper number.

Even when this kind of number record becomes highly abbreviated or systematized, we are not necessarily led to the concept of zero. For example, suppose people wrote '6E4' to indicate six pebbles in the third bowl, an empty second bowl and four pebbles in the first. Our symbol 'E' can still be read as something rather like a punctuation mark, and not as something that is equivalent in kind to the other numerals '6' and '4'. Indeed, the same comment could be made about modern decimal points. They appear in numerical expressions alongside the digits 0 to 9, but no one thinks the decimal point is a number.

Returning to the Ancient Sumerians, it is remarkable to note that even though they developed positional notation, and (somewhat later) a symbol comprised of two diagonal lines that worked rather like the modern '0', their system did not spread to other civilizations. I say that this is remarkable because compared with Roman numerals (say), positional numeration is awesomely convenient. For example, imagine trying to multiply LXXVII by CXI, and then think how much easier it is to multiply 87 by 111. The crucial point is that Roman numerals, like the English words for numbers, have very little connection to the workings of an abacus. That is to say, adding groups of ten is entirely akin to adding groups of a thousand, but this mathematical truth is not well reflected in the English language or in Roman numerals.

In particular, if you want to multiply a pair of numbers that are represented as Roman numerals, you cannot simply break down the problem by multiplying one digit at a time. In contrast, positional notation naturally leads us to the observation that $87 \times 111 = 8{,}700 + 870 + 87$. For similar reasons, addition and subtraction are much easier with positional notation, as like an abacus, the notation shows us that we can add one numeral at a time. Furthermore, positional notation makes it much easier to write large numbers, as the same numerals are used for numbers of every size. Indeed, in Roman numerals, even a number as small as one million can be represented only by an unwieldy and easily misread string of one thousand Ms.

Alexander the Great must have encountered important records that were written with positional notation, and so presumably the idea made its way back to Greece, but for whatever reason it didn't catch on, and was soon forgotten. In any case, our modern number system doesn't just require the basic idea of positional notation. Negative numbers are another essential concept, as it is hard to believe that any civilization could even invent the number zero without first using negative numbers. The use of negative numbers started sometime around 200 BC, when the Chinese started using red sticks on a counting board to indicate a credit, while black sticks indicated a debt. Ancient textbooks for Chinese civil servants included instructions that said 'If your board contains red sticks and black sticks, remove an equal number of both.' In other words, they understood that equal-sized positive and negative numbers cancel each other out.

Of course, using the integers to represent both credits and debts is not the same thing as having a single number system that incorporates both positive and negative numbers, even if it is well understood that subtracting a credit of n from someone's ledger is equivalent to adding a debt of n, while subtracting a debt of n is equivalent to adding a credit of n. My point is that to have a proper system of positive and negative integers, we need to be able to add, subtract and multiply any pair of integers to get another integer, and we cannot have a number system of this type without first conceptualizing 'nothing' as a number. This conceptual leap was a key event in the history of ideas, and it took place in India.

Nobody knows who invented the symbol zero, but there is little doubt that such a symbol was in existence in parts of India as far back as 500 BC, although back then it was not widely used. At first this symbol acted as a kind of placeholder, which represented an empty column on a counting board. It was a brilliant and convenient addition to the local positional system, but for a thousand years or so, the zero symbol did not really represent a number as such. The first truly definitive remark concerning the *number* zero was made by a mathematician and astronomer called Brahmagupta of Gujarat (*c.* 598–670). He said that zero is the number you get when you combine a credit and a debt of equal value (i.e. he observed that '$n - n = 0$'). He also stated that one plus zero is one, one minus zero is one, and one times zero is zero. These axioms were crucially important, because they put zero into the same conceptual domain as the other numbers. For example, we might say six minus five leaves a credit of one, five minus six leaves a debt of one, while six minus six leaves zero, viewing all three statements as being of a similar kind.

Brahmagupta's name for the number zero was '*sunya*', which means void or empty, and long before the time of Brahmagupta, the word *sunya* was used to describe a column on a counting board that did not contain any markers. *Sunya* could also be translated as 'space', since Ancient Indian architects noted that they did not design walls so much as the space (*sunya*) between the walls. The various uses of the word *sunya* before the invention of the number zero are telling, and it is no coincidence that the mathematicians who developed the concept of zero belonged to the Vedic tradition, where discussions of the qualities of the void or 'emptiness' had been taken seriously for thousands of years. Indeed, I think it is significant that where Christians, Muslims and Jews tend to associate the divine with the infinite, Hindus, Buddhists and Jains more frequently associate the divine with a conception of emptiness, as the very essence of nirvana is the absence of desire.

When the Arabs adopted Indian numeration in the tenth century, the word *sunya* was translated to the Arabic word for empty, *sifr*. The English words cipher and decipher are derived from the word *sifr*, as when the modern, decimal system was introduced to Europe, many people needed to decipher such numbers into Roman numerals before they could understand them. The word zero comes from the same root, as in the early thirteenth century the word *sifr* was Latinized into zephirum, which later developed into the word zero.

Al-Khwarizmi and the Science of Equations

The history of mathematics would be very different if it were not for the influence of people from Iran and Iraq. Before and after the rise of Islam, the ruling elite in that part of the world took a serious and scholarly interest in the intellectual developments of their neighbours, which included the exceptionally sophisticated mathematicians of Constantinople, Alexandria, India and China. Their great legacy of exact science was based on a remarkably broad assessment of humanity's mathematical knowledge, with scholars searching far and wide in order to further their own intellectual interests. For example, translations were made of the *Siddhantas* (important Hindu texts concerned with mathematics and astronomy), and it is highly probable that there were scholars of Brahmagupta's masterpiece the *Brâhmasphutasiddhânta*, which was the world's most advanced number-theoretic text when it was written back in AD 628. As well as introducing zero and rules for manipulating negative and positive numbers, this influential work introduced a fully general form of the quadratic equation, which we still use today. Most advanced of all, Brahmagupta's masterpiece also contained a technique for solving certain 'Pell equations', such as $x^2 - 92y^2 = 1$.

Arab scholars studied widely, acquiring mathematical knowledge that was utterly unknown to Europeans. They also preserved and studied the works of Ancient Greece, maintaining traditions that Europe had lost. As well as performing the invaluable service of pooling and consolidating the world's mathematical knowledge, the Arabs also developed their own innovations, and crucially they introduced Europeans to the many benefits of the decimal position system. The most famous and influential of the

Baghdadi mathematicians was Abu Ja'far Mohammed ibn Musa al-Khwarizmi (*c.* AD 780–850), or al-Khwarizmi for short. The word algorithm is a Latinization of his name, and the word algebra is derived from the title of his most famous book, *Al-Jabr W'al Mûqabalah*, or 'Calculation by Restoration and Reduction'.

It is hard to overstate the importance of the basic ways that scientists and mathematicians manipulate the symbols in equations to produce other, equally valid equations. For example, following al-Khwarizmi, we learn that we can collect a number of terms together on one side of the equals sign, or multiply both sides of an equation by any convenient number. Although al-Khwarizmi fully deserves his fame, it is only fair to stress that his general methods for finding unknown quantities were developments based on a number of ancient traditions. In particular, some people refer to Diophantus (*c.* 210–294) as the father of algebra. Very little is known about Diophantus, but he lived in Alexandria in the second century AD, and famously worked to identify integer solutions to various kinds of polynomial equation. That is to say, he studied equations involving the addition and multiplication of integers together with some unknown quantity, such as $2x^2 + 10 = 60$. His distinctive contribution was not only to use a single symbol to represent an unknown quantity x (as earlier authors are thought to have done): he also used a symbol for the square of an unknown quantity, as well as the cube of an unknown quantity.

The mathematics of Diophantus was highly original because he was the first person to relate unknown quantities in equations that can have an infinite number of solutions. However, the equations that he studied were always related to geometric problems, and in that context his work did not naturally lead to a radically new branch of maths. The fundamental ways that we can legitimately rearrange or manipulate equations were essentially identified by al-Khwarizmi, and his *Al-Jabr* was accepted for centuries as the definitive text on the science of handling arithmetic problems involving unknown quantities.

The crucial difference between the Ancient Greeks and the Arabs was that the Arabs considered linear equations, quadratic equations, cubic equations and so on to be distinct and fundamental categories of problem. The Greeks treated such problems as part of their assault on geometry, while the Arabs saw that the language of equations can be applied to a whole range of mathematical areas of interest, including geometry and number theory. They therefore recognized that finding the legitimate ways to manipulate equations is a worthy enterprise in its own right. For example, although al-Khwarizmi worked with word equations, and not a symbol like the modern x, he effectively stated that an equation of the form $x=40-4x$ can be rewritten as $5x=40$. Furthermore, he made it clear that collecting terms together was a fundamental and general principle, as his principles of 'Reduction' and 'Restoration' effectively tell us that we can do whatever we like to one side on an equals sign so long as we do the exact same thing to the other side as well. This basic technique can be used in all manner of situations, as, for example, we can divide both sides by 5 in order to reduce the equation $5x=40$ to the simpler form $x=8$.

The stated purpose of al-Khwarizmi's celebrated work was to show 'what is easiest and most useful in Arithmetic, such as men constantly require in cases of inheritance, legacies, partition, lawsuits, and trade, and in all their dealings with one another, or where the measuring of lands, the digging of canals, geometrical computations, and other objects of various sorts and kinds are concerned'. Al-Khwarizmi clearly understood that the principles he was explaining were exceedingly general, holding true across all of the mathematical subject matters considered by the ancients. Indeed, al-Khwarizmi rather charmingly remarked on this unity of mathematics with the simple comment, 'When I consider what people generally want in calculating, I found that it is always a number.'

The notion that techniques for handling equations might constitute an independent branch of mathematics was an extremely important development, which the Arabs were uniquely placed to make. As John Stillwell wrote in his classic *Mathematics and its History*:

> 'In Indian mathematics, algebra was inseparable from number theory and elementary arithmetic. In Greek mathematics, algebra was hidden by geometry. Other possible sources of algebra, Babylonia and China, were lost or cut off from the West until it was too late for them to be influential. Arabic mathematics developed at the right time and place to absorb both the geometry of the West and the algebra of the East and to recognize algebra as a separate field with its own methods. The concept of algebra that emerged – the theory of polynomial equations – proved its worth by holding firm for 1000 years. Only in the nineteenth century did algebra grow beyond the bounds of the theory of equations, and this was at a time when most fields of mathematics were outgrowing their established habitats.'

Algebra and Medieval Europe

In Europe, the basic syllabus of early medieval science could be found in a fairly small number of famous books, each of which was written in Latin. To surpass their peers in mathematics and science, scholars turned to Ancient Greek texts, and the science of the Arab world. European scholars were slow in learning from the Arabs, but it seems clear that their influence was ultimately pivotal. A particularly early example of European Christians learning Arabic science can be found in the truly remarkable figure of

Gerbert of Aurillac (*c.* 946–1003). Gerbert was born in France, but in 967 he moved close to Barcelona, studying under the direction of Atto, Bishop of Vic. There was conflict in Spain between the Christians and the Muslims of al-Andalus, and in the face of a Christian defeat, Atto was charged with delivering a request for a ceasefire.

Atto was received as an honoured guest, and soon found himself mesmerized by the palaces in Cordoba. His brilliant student Gerbert shared his fascination with the Arabs, whom he greatly admired for their knowledge of mathematics, astronomy and science. A prolific scholar and a gifted teacher, Gerbert is credited with reintroducing Europeans to the abacus and the armillary sphere: a kind of visual aid for teaching mathematics and astronomy that had been lost to Europe since the end of the Greco-Roman era. Remarkably, Gerbert's abacus incorporated Arabic numerals (many centuries before modern, decimal notation was used elsewhere in Europe), and it was rumoured that when he was young, Gerbert would sneak out from the monastery at night to study under the guidance of the Arabs.

Gerbert was appointed as a teacher to the Holy Roman Emperor Otto II, and in 999 he reached the very head of the Western Church, becoming the first French pope (Pope Sylvester II). On 31 December he celebrated a solemn mass as the people of Rome trembled before the impending apocalypse, but contrary to expectations the sun rose on the year 1000, and Gerbert remained as pope until his death on 12 May 1003. In those turbulent times knowledge of advanced science was rare, and for several centuries very few Europeans were familiar with decimal numbers. The spread of knowledge was much slower in those times, but by the twelfth century Arabic texts were being translated into Latin and distributed across Europe.

Perhaps surprisingly, there was a gap of several centuries between the introduction of decimal numbers and their widespread use by merchants and administrators. The first widely read book on decimal numbers was Leonardo Fibonacci's *Liber Abaci*, written in 1202, but the popular response to this innovation was largely hostile. Indeed, in 1299 the city of Florence went so far as to make the use of decimal numbers a criminal offence! Generations of the general public suspiciously dismissed them as a form of financial trickery, and it was not until the end of the fourteenth century that ever increasing numbers of people abandoned Roman numerals for the superior decimal system. For example, the largest and most powerful bank in Europe, the Medici Bank, did not switch their account books to the decimal system until 1439 77m^2.

Decimal numeration is a very powerful system, and the introduction of algebra also had a massive impact on the development of European maths and science. We have seen that even in Ancient Babylon, people could answer questions such as: 'A rectangle has an area of 77m^2, and one side is 4m longer than the other. How long are the two sides?' In modern notation, we would write $x(x+4)=77$, or $x^2+4x-77=0$. My point is that the form of writing known as an equation is a relatively modern innovation, though it can be used to rewrite ancient problems, and this powerful innovation has Arabic roots. Indeed, in the modern world, our uses and misuses of equations have become central to our attempts to understand the world.

A critical point is that we can use the very same equations to summarize all manner of analogous things, not just the geometric facts that the first 'equations' referred to. As everybody knows, modern science is filled with equations. Furthermore, Jacob Klein and other scholars have convincingly argued that using equations to study the integers has led to a subtle shift and abstract expansion of our conception of number. For example, we know that every integer or counting number is either even or odd. If the integer N is even, it can be written in the form $N=2x$, (where x is simply another integer). Similarly, if N

is odd it can be written in the form $N=2x+1$. More generally, given any positive integer M, any integer N must be expressible in one (and only one) of the following forms:

$$N=Mx \text{ (where } x \text{ is some integer), or } N=Mx+1,$$

$$\text{or } N=Mx+2, \text{ or } N=Mx+3, \ldots, \text{ or } N=Mx+M-1.$$

In other words, every integer N is either a multiple of M, or it is one bigger than a multiple of M (so dividing N by M leaves a remainder of one), or it is two bigger than a multiple of M (so dividing N by M leaves a remainder of two), and so on up to the largest possible remainder, $M-1$.

This kind of observation was explored by the eminent lawyer and mathematician François Viète (1540–1603). Viète wanted to understand the different kinds of pattern that are produced when we divide various integers and examine the remainders. He carried out this investigation in a systematic fashion, introducing an algebraic notation that is much like the one used today. More specifically, if N divided by M leaves a remainder of n, we say that N is equivalent to n modulo M, and write $N \equiv n \pmod{M}$. For example, $3 \equiv 1 \pmod{2}$, because three divided by two leaves a remainder of one. Similarly, $10 \equiv 1 \pmod{3}$, because ten divided by three leaves a remainder of one.

Now suppose that we have three integers A, B and M, such that A divided by M gives a remainder a, while B divided by M gives a remainder b. A profound question to ask is whether it is always true that $A+B \equiv a+b$ (mod M), and $A \times B \equiv a \times b \pmod{M}$. As an example, suppose that $M=10$, $A=53$ and $B=12$. Dividing by 10 and taking the remainder is equivalent to looking at the final digit of the numbers A and B. If you want to know the final digit of 53×10, you simply calculate 3×2. The larger digits only affect the first digits of the answer – they cannot affect the final digit.

This observation suggests that if we divide by M and take the remainder, then do our sums with a and b, we must get exactly the same answer as when we do our sums with the numbers A and B and then divide by M and take the remainder. Indeed, this very general fact is easy to prove using algebra. We simply write A in the form $a+xM$ and B in the form $b+yM$, where x and y are integers, and a and b are non-negative integers smaller than M. We now note that:

$$(a+xM)+(b+yM)=a+b+(x+y)M, \text{ and}$$

$$(a+xM)\times(b+yM)=a\times b+(bx+ay+xyM)M.$$

This proves that if $A\equiv a$ (mod M) and $A\equiv b$ (mod M), then $A+B\equiv a+b$ (mod M) and $A\times B\equiv a\times b$ (mod M). Hence we can prove statements like 'odd times odd is always odd' *without needing to consider specific integer cases.* We simply note that an integer is odd if and only if it can be written in the form $1+2x$, and that:

$$(1+2x)\times(1+2y)=1+2(x+2xy+y).$$

In other words, we have proved that $1\times1\equiv1$ (mod 2). Similarly, it is very easy to show that $1+1\equiv0$ (mod 2).

The crucial point is that following the precedent set by François Viète, one can conceive of settling facts such as 'odd times odd is odd' by exploiting a purely algebraic system. By doing this we find ourselves systematically dealing in 'species' of number (such as even or odd), without needing to refer to the counted collections that were the ultimate point of reference for ancient mathematicians. Such equations correctly summarize the individual facts where x and y are some definite, counted out pair of integers, but they also state a more general, abstract truth.

We can read the equation $(1+2x)\times(+2y)=1+2(x+2xy+y)$ even when we interpret the symbols x and y as inherently variable. In other words, we can read $2x+1$ as our way of symbolically representing the general form called 'odd'. What is more, the expression $2x+1$ is seen as having an essentially numerical character precisely because it can serve as an item within a computational system. That is to say, we know how to add and multiply expressions such as $2x+1$ and $2y+1$, and doing these kinds of sum does not require us to ever substitute any specific integer values for the variables x and y.

Fermat's Little Theorem

By the seventeenth century European mathematicians were taking full advantage of the basic techniques of algebra. By working with symbols for unknown quantities and by studying general forms of equation, mathematicians could now prove new kinds of number-theoretic results. A particularly beautiful example of this relatively modern form of number theory is Fermat's Little Theorem: one of the many important results proved by Pierre de Fermat (1601–1665). As Fermat stated in a letter from 1640, if we take any prime number p and any integer n, we can be certain that:

$$n^p \equiv n \pmod{p}.$$

In the next section we will see how this theorem underpins the construction of the mathematical padlocks we use to protect electronic business transactions. First, I want to sketch a proof of remarkable theorem.

The fundamental observation behind Fermat's Little Theorem concerns the way that we multiply brackets filled with additions. For example, $a(b+c)=ab+ac$, and $(a+b)(c+d)=ac+ad+bc+bd$. In general, when we multiply out a sequence of bracketed terms, we take one thing from each bracket and multiply them together to form one of the terms on the right-hand side. Once we have taken one term from each bracket every different way we can, we add up all the resulting terms to get our final answer.

In particular, consider how we would rewrite the following expression without using brackets:

$$(x_1 + x_2 + ... + x_n)^2 = (x_1 + x_2 + ... + x_n) \times (x_1 + x_2 + ... + x_n).$$

Taking the first term from each bracket and multiplying them together gives us x_1^2. Similarly, taking the second term from each bracket and multiplying them together gives us x_2^2, taking the third term from every bracket and multiplying them together gives us x_3^2, and so on. When, for example, we take the first term from the first bracket and the second term from the second bracket, we get x_1x_2. If we take the second term from the first bracket and the first term from the second bracket we also get $x_2x_1 = x_1x_2$. Because we can pick the term x_1 from either the first or the second bracket, our final expression will contain the term $2x_1x_2$. More generally, our final expression will have the form

$$x_1^2 + x_2 + ... + x_n^2 + 2f(x_1, x_2, ..., x_n).$$

where $f(x_1, x_2, ..., x_n)$ is a polynomial expression involving the variables x_1 to x_n. Now let's consider how we would rewrite the following expression without using brackets:

$$(x_1 + x_2 + ... + x_n)^p = (x_1 + x_2 + ... + x_n) \times ... \times (x_1 + x_2 + ... + x_n).$$

Taking the first term from every bracket and multiplying them together gives us x_1^p, taking the second term from every bracket and multiplying them together gives us x_2^p, and so on. When, for example, we take the first term from the first bracket and the second term from every other bracket, we get the expression $x_1x_2^{p-1}$. Similarly, if we take the first term from the second bracket, and the second term from every other bracket, we also get $x_2x_1x_2^{p-2} = x_1x_2^{p-1}$. Because we can pick the term x_1 from any of the p different brackets and still get the same answer, our final expression must contain the term $px_1x_2^{p-1}$. More generally, we must be able to write our final expression in the form $x_1^p + x_2^p + ... + x_n^p + pf(x_1, x_2, ..., x_n)$, where $f(x_1, x_2, ..., x_n)$. is a polynomial expression involving the variables x_1 to x_n.

We have just argued that

$$(x_1 + x_2 + ... + x_n)^p = x_1^p + x_2^p + ... + x_n^p + pf(x_1, x_2, ..., x_n),$$

which nearly completes the proof of Fermat's Little Theorem. All we need to do is consider the case where $x_1 = x_2 = ... = x_n = 1$. In this case we have

$$(1 + 1 + ... + 1)^p = 1^p + 1^p + ... + 1^p + pf(1, 1, ..., 1)$$
$$\text{and so } n^p = n + pf(1, 1, ..., 1).$$

Since f is a polynomial equation with integer coefficients, it follows that $f(1, 1, ..., 1)$ must be an integer. Hence, we have proved that for every integer n and every prime number p, n^p must equal n plus some multiple of p. In other words, for every integer n and every prime number p, n^p is equivalent to n modulo p. If we divide both sides by n we can see that $n^{p-1} \equiv 1 \pmod{p}$, and as we shall see in the following section, this fact lies at the heart of Internet security.

How to Make a Mathematical Padlock

For every composite number n, there is a very simple proof of the fact that n is composite. All we require are two numbers (a and b), such that $a \times b = n$. In contrast to this short and simple proof that the number n is composite, a proof that a given number is prime is (necessarily?) large. That is to say, every known method for factorizing numbers involves a huge amount of brute calculation, and showing that a number has no factors is similarly time-consuming.

Some astronomically large numbers have been shown to be prime. In fact, the largest known primes are much bigger than astronomical, as the largest known prime is much, much bigger than our best estimates for the number of atoms in the universe. Strangely enough, large prime numbers have a financial value, because they are essential for the construction of the mathematical padlocks that are used to protect electronic business transactions.

These mathematical padlocks are rules that convert an input into an output. Corporate entities publish these rules (i.e. they freely distribute padlocks), and someone who wants to encode the message 'I will buy a million shares' simply feeds their message into the rule, which converts their message into some large number C. The person then transmits the number C, and at the other end this number is fed into a secret rule (the key to the padlock), which converts the encoded message back into ordinary text.

Someone who eavesdrops on these transmissions can, in principle, work out what any encoded message says without being told the secret rule. For example, they could try encoding every possible statement in turn, and compare the results with the encoded message C (that is, the scrambled version of the message that they want to read). If one of our hacker's made-up messages happens to encode to exactly the number C, that tells our hacker that their made-up message is actually the same as the unscrambled version of the message that they want to translate. Of course, encrypting every possible message that a person might have sent is not a practical way to eavesdrop, and it is believed that successfully cracking these codes requires an astronomical number of calculations (unless you know the secret rule, of course). That is to say, finding the key by examining the padlock is practically impossible.

One standard form of mathematical padlock is called the RSA system, after the trio of mathematicians and computer scientists who first published the algorithm.[1] This kind of mathematical padlock takes a fixed number of binary digits as its input. Any sequence of binary digits can be interpreted as an integer M. Given such an integer M, the output of the system is

$$C \equiv M^e \pmod{n},$$

where e and n are fixed integers, specified by the people who are handing out the padlock. Most real-life RSA cryptosystems use values for e and n with over a thousand digits. Because we are doing these calculations modulo n, it is actually quite simple for a modern-day computer to calculate the correct output. However, doing this procedure 'backwards' is riddled with difficulties and very, very time-consuming. It is somewhat analogous to trying to piece together something that has gone through a shredder – even the person who put the message in can't undo it. By way of contrast, the secret rule is very simple: we just take the encoded message C, and calculate $C^d \pmod{n}$.

The whole point of this system is that $C^d = (M^e)^d = M^{ed} \pmod{n}$. To make this system work, all we require are three numbers, e, d and n, such that for every message M, we have $M \equiv M^{ed} \pmod{n}$. That way, if we run a message through the padlock and then through the key, we get back what we started with.

Given two prime numbers p and q, we can find the

1 The RSA system is named after Ron Rivest, Adi Shamir and Leonard Adleman, who devised the algorithm in 1977. In fact, an equivalent scheme was described in 1973 by the British mathematician and cryptographer Clifford Cocks, but since Cocks was employed by the Government Communications Headquarters at the time, his work was classified top-secret, and did not enter the public domain until 1998.

required numbers by putting $n = p \times q$. The integers e and d must be smaller than $(p - 1)$ times $(q - 1)$, and it is also essential that $ed \equiv 1 \pmod{(p-1)(q-1)}$. In other words, there must be some integer t such that $ed = 1 + t(p\text{-}1)(q\text{-}1)$. Given any two prime numbers p and q, it is actually quite easy to find a pair of integers e and d with the desired property.

Now, given any prime number p, Fermat's Little Theorem tells us that:

$$M^{p-1} \equiv 1 \pmod{p}.$$

Raising both sides of this equation to the power s gives us the following relationship:

$$M^{s(p-1)} \equiv 1 \pmod{p}, \text{ for every integer } s.$$

In particular, $M^{t(q-1)(p-1)} = M^{ed-1} \equiv 1 \pmod{\mathrm{p}}$. In other words, $M^{ed-1} - 1$ is a multiple of p. A similar argument shows that $M^{ed-1} - 1$ is also a multiple of q. Since p and q are both prime numbers, this tells us that $M^{ed-1} - 1$ must be a multiple of $n = p \times q$. In other words, $M^{ed-1} \equiv 1 \pmod{n}$. By multiplying both sides of this equation by M, we get the result we were looking for, namely that:

$$M^{ed} \equiv M \pmod{n}.$$

To recap, raising the message M to the power e gives us the coded message $C \equiv M^e \pmod{n}$. Raising the coded message C to the power d gives us $M^{ed} \equiv M \pmod{n}$, which means that our secret rule correctly decodes messages. It is theoretically possible to calculate the secret code number d by using the public padlock numbers n and e. The only difficult bit is factorizing the integer n, because everything else can be calculated relatively quickly.

This means that the security of the world's financial markets rests on the assumption that factorizing vast numbers is very difficult (or to be more accurate, that cracking RSA is very difficult). It is reasonable to assume that this is so, but it really is an assumption. Although mathematicians can prove that huge classes of problems are equally difficult (in a particular well-defined sense), there are serious difficulties in proving that some problems are harder than others! We can see that a particular method for solving one problem is more time-consuming than some particular method for solving a second problem, but that is not the same thing at all. There is literally a million-dollar reward for the person who can prove an inherent distinction of difficulty, where we show that any possible method for solving one class of problem necessarily takes longer than some given method for solving a second class of problem.

Time and again, great progress has been made when one branch of mathematics has been brought to bear on another. For example, Fermat's Little Theorem and modern mathematical padlocks are both examples of algebraic number theory. As we shall see in the following chapter, the basic techniques of algebra were also pivotal in enabling the development of algebraic geometry. Without this new approach to the ancient discipline of geometry, modern science could not have developed, and we would not have scientific equations as we know them today.

Chapter 5:
MECHANICS AND THE CALCULUS

> 'The heart of mathematics consists of concrete examples and concrete problems. Big general theories are usually afterthoughts based on small but profound insights; the insights themselves come from concrete special cases.'
>
> Paul Halmos, 1916–2006

The Origins of Analysis

In this chapter I will examine the origins of calculus, and the co-evolution of two very different branches of mathematical science: analysis and mechanics. Roughly speaking, mathematical analysis is the study of infinite, mathematical sequences, while mechanics is the study of moving objects. As we shall see, the science of mechanics played a critical role in the development of analysis, particularly the infinitesimal calculus. On the other hand, many ideas relating to infinite sequences are extremely old, predating our ability to mathematically analyze motion. Indeed, we have already encountered the oldest analytic concepts, such as the intuitive idea of generating an infinite sequence, and finding the limit case.

For example, one of Archimedes' most elegant proofs

concerns the area under a parabola. Imagine, for example, that we have fired a projectile up into the air from a horizontal firing range. The projectile will follow a parabolic path, and we can draw a triangle between the firing point, the highest point in the arc, and the point where the projectile lands. It was obvious to Archimedes that if this triangle has an area A, the parabola must have an area greater than A (precisely because the triangle fits inside the parabola). As a small part of the parabola was left uncovered, he added two more triangles, and he did this in a clever way so he could prove that those two triangles must have a total area of $\frac{A}{4}$. After adding those triangles an even smaller part of the parabola remained uncovered, so he added a further four triangles with a total area of $\frac{A}{4^2}$. Furthermore, he made it clear that you can continue adding smaller and smaller triangles indefinitely, where at each stage the total area of the additional triangles was one quarter of the previous stage.

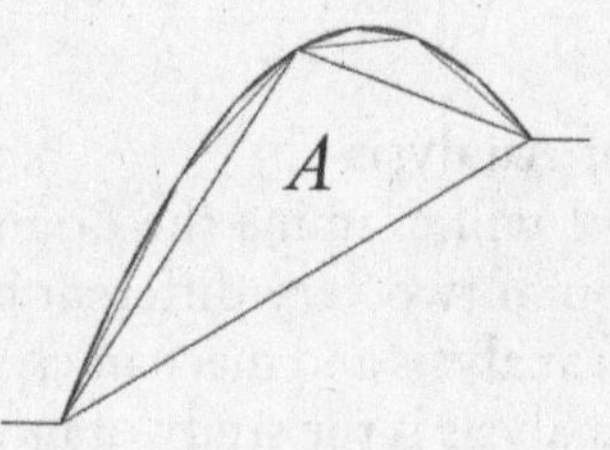

Archimedes could prove that the triangle marked A has four times the area of the two triangles directly above A. He could also prove that those two triangles have four times the area of the four triangles directly above them, and so on.

Up until the nineteenth century, the area under a curve was taken to be a *given* quantity. In other words, Archimedes simply assumed that a truncated parabola defines a definite quantity, namely the area underneath it. Not every curve has a well-defined area underneath it, but up until the time of Carl Friedrich Gauss (1777–1855), mathematicians only studied curves that had well-defined areas underneath them. Because it was intuitively obvious that the shapes in question had a definite area, mathematicians were justifiably confident when using an infinite sequence to *calculate* that area. My point is that we don't need to worry about whether or not $A(1+\frac{1}{4}+\frac{1}{4^2}+\frac{1}{4^3}+\ldots)$ *defines* one specific quantity. Our geometric intuition is enough to convince us of the fact that the area under a particular parabola is equal to one particular quantity, and this leads us to accept the fact that $1+\frac{1}{4}+\frac{1}{4^2}+\frac{1}{4^3}+\ldots$ picks out exactly one real number. Given that this limit case is indeed a number, we can safely make the following argument:

Since $x = 1 + \frac{1}{4} + \frac{1}{4^2} + \frac{1}{4^3} + \ldots$

it follows that $4x = 4 + 1 + \frac{1}{4} + \frac{1}{4^2} + \frac{1}{4^3} + \ldots$

Subtracting the first equation from the second tells us that $3x = 4$, which means that $x = \frac{4}{3}$.

Archimedes could see that $1 + \frac{1}{4} + \frac{1}{4^2} + \frac{1}{4^3} + \ldots = \frac{4}{3}$, but he was understandably wary of any argument that built on the idea of an infinite sequence. He therefore established the validity of his answer by using Proof by Exhaustion. In other words, he rigorously proved that the area under a parabola cannot possibly be smaller than $\frac{4A}{3}$, and it cannot be larger than $\frac{4A}{3}$, so it must equal $\frac{4A}{3}$. Although Archimedes' proof does not depend on the properties of infinite sequences, the notion of slicing a shape into thinner and thinner pieces is at least twenty-two centuries old. We know that this is so because Archimedes explained this ingenious idea in a work entitled *The Method*.

This remarkable text had been lost since the Middle Ages, so the modern inventors of calculus certainly hadn't read it. Indeed, until very recently, everything we knew about the mathematics of Archimedes was ultimately derived from just two manuscripts (codices A and B). However, in 1906 the Danish philologist J. L. Heiberg discovered a third ancient source (codex C), which included *The Method*. This particular collection of works of Archimedes had been overlooked for so many generations because it had been copied onto parchment in the tenth century, but a couple of hundred years later the precious material had been washed down, rebound, and inscribed with prayers. The original text was almost invisible, but in the Monastery of the Holy Sepulchre, Istanbul, it lay waiting to be discovered. Fortunately for us, parts of the original text could still be seen, and thanks to x-ray technology almost all of it has now been restored.

In the course of *The Method* Archimedes reveals that many of his deepest and most original results were found by using questionable infinitary arguments, and only later did he go on to prove his results using the rigorous Proof by Exhaustion. In other words, Archimedes' constructed arguments using infinite sequences, but as the logical foundations of such arguments were unclear, he refused to rely on them while stating his proofs. Nevertheless, *The Method* tells us that 'It is of course easier to supply the proof when we have previously acquired some knowledge of the questions by the method, than it is to find it without any previous knowledge.' In other words, it is easier to write a rigorous proof when you already know the right answer.

Since the 1970s scholars have also known about another form of ancient analysis, which was hidden in the astronomy of medieval India. In particular, the Kerala region in the south-west of India produced one of the great mathematical visionaries: Madhava of Sangamagrama (*c.* 1350–1425). All the mathematical writings of Madhava have been lost, though some of his texts on astronomy have survived. We know about his brilliant mathematical work through reports from later centuries, most of which were written in the regional language of Malayalam. Trigonometric problems routinely arise while plotting the course of the stars, and Madhava discovered and extended rules for solving these kinds of problem. Crucially, his technique involved infinite sequences. For example, he knew that:

$$\pi/4 = 1 - 1/3 + 1/5 - 1/7 + \dots$$

Madhava needed numerical answers to put in the ledgers of astronomy, and, most impressively, when he used an infinite sequence to generate an approximate answer, he commented on the size of his margin of error. For example, in the case of finding $\pi/4$, he knew that using n terms gives an approximation that is within a margin of error of $1/2n$. In a remarkable passage describing the construction of this kind of sequence, the mathematician-astronomer Jyesthadeva (*c.* 1500–1575) states that care must be taken 'otherwise the correcting term [or margin of error] will not tend to the vanishing magnitude'. This line indicates that Madhava's work prefigured the modern definition of a limit case by four or five hundred years.

The crucial point is that some sequences do not have a limit case, and we cannot always rely on geometric intuition to reassure us that a limit must exist. For example, the sequence 1, 1 − 1, 1 − 1 + 1, 1 − 1 + 1 − 1, ... is fundamentally different to 1, $1 - 1/3$, $1 - 1/3 + 1/5$, ... The first sequence hops between 1 and 0, and never settles on a limit. The second sequence narrows in on one particular number, and the margin of error becomes arbitrarily small.

For this reason we say that $1 - \frac{1}{3} + \frac{1}{5} - \frac{1}{7} + \ldots$ is a convergent series, which defines a real number. In contrast, the string of symbols $1 - 1 + 1 - 1 + \ldots$ is said to be divergent, and this second sequence does not define a real number.

Modern mathematicians say that by definition, a sequence $x_1, x_2, x_3, \ldots$ converges to a limit L if and only if:

> For every positive number δ, there is some number n such that every term $x_n, x_{n+1}, \ldots$ is bigger than $L - \delta$ and smaller than $L + \delta$.

In other words, a point L is the limit of a sequence if and only if *any* 'target region' containing L contains all but a finite number of the points in our sequence. As we shall see in a later chapter, it is crucial that our definition uses the logical words 'and', 'every' and 'some', because this kind of vocabulary can support various forms of logical deduction. Furthermore, we should note that although certain analytic ideas have ancient origins, this logical, axiomatic foundation for modern analysis did not arise until the early nineteenth century. Before that time the scope of calculus was not fully apparent, and in the hands of less expert men, many more mistakes could have been made.

Measuring the World

Most branches of mathematics have been influenced by empirical science, but the mathematical ideas presented in this chapter have a particularly close relationship with the study of moving objects. It is a fundamental fact that the study of moving objects requires the measurement of time. For example, countless generations have tried to calculate the time it takes for the planets to complete their orbits across the heavens, and this kind of inquiry was critical to the development of modern, mathematical science. In order to convey something of the intellectual traditions that preceded Newtonian mechanics, let's consider the work of two masters of measurement: al-Biruni and Galileo Galilei.

The Persian scholar Abu Rayhan Muhammad ibn Ahmad al-Biruni (973–1048) was an exceptionally advanced scholar, and we can use him to gain an impression of the best of medieval science. A great polymath, al-Biruni was right at the cutting edge of mathematics, science and the humanities. His most famous book was *Tarikh Al-Hind* (or 'History of India'), a remarkable account of the ritual, mathematical and astronomical knowledge of Hindu India. He was also an expert on Greek mathematics and science, but unlike many of his contemporaries, he was prepared to challenge Aristotle. For example, it was widely believed that cooling an object makes it shrink, but al-Biruni challenged this assumption, noting that a glass full of water will crack when the water freezes. An expert writer of lists, he was also the first person to define explicitly the concept of relative density, compiling tables of the empirically determined densities of various gemstones and metals.

In Europe it took longer for mathematical physics to develop, and until the time of Galileo Galilei (1564–1642), Ancient Greek texts remained the pinnacle of science. The route to further progress required a combination of theoretical and empirical work, but the defining spirit of the scientific revolution could be summarized by Galileo's powerful slogan: 'Measure what is measurable, and make measurable what is not.' Galileo was appointed to the chair of mathematics in Pisa, and over the course of a long and varied career, his careful measurements of moving bodies led to a deeply mathematical understanding of motion. He gave an accurate, mathematical account of the swinging of a pendulum, and he was the first person to argue that projectiles should follow a parabolic path. He was also one of the first people to explore the night sky with a telescope, identifying the moons of Jupiter, and observing that the haze of the Milky Way is comprised of countless stars.

Because of his many achievements, and the fact that he explicitly elevated mathematical reasoning and empirical data as the ultimate guides to nature, Galileo is sometimes called the father of modern science. One of his most impressive insights was a claim that was essential to the development of Newtonian mechanics. That is to say, Galileo famously stated that under the force of gravity alone, all objects accelerate at the same rate. In other words, it is only air resistance that stops stones and feathers from falling side by side. Galileo argued that this statement must be true, not because of the results of some physical experiment (though he was certainly a keen experimentalist), but because of a mathematical thought experiment.

This conceptual experiment began with the assumption that heavy bodies fall faster than lighter ones, as Aristotle claimed. If this were a fundamental truth (as most people assumed), Galileo wondered what would happen if you dropped a heavy cannon ball attached to a lighter musket ball. According to Aristotle's theory, the musket ball would trail behind the cannon ball because of its lighter weight. This would mean that if the two balls were attached to one another, the slower moving, lighter ball would slow down the faster moving, heavier ball, so the combination of two balls would fall more slowly than the cannon ball on its own.

On the other hand, the combined weight of a musket ball and a cannon ball is greater than the weight of a cannon ball on its own. Hence Aristotle's theory also implies that the combination of two balls should fall faster than the cannon ball on its own. This analysis is devastating for Aristotle's account of gravity. It cannot possibly be the case that a heavy ball falls both faster and slower than the even heavier system of a heavy ball attached to a light ball. Now, given that Aristotle's theory of falling bodies has been shown to be unworkable, what should we put in its place? The same line of reasoning would destroy the absurd notion that lighter objects fall faster than heavier ones, leaving only one readily conceivable option: the effect of earth's gravity is to make all objects fall at the same rate.

The Age of Clocks

Ancient accounts of motion or change were largely focused on classifying the form or 'cosmic purpose' of the various kinds of change that people had observed. Properly quantifying how quickly objects fall (for example) was a kind of science that did not develop until the Renaissance, with Galileo being a key, transitional figure. It is worth stressing that mechanics is essentially concerned with the movement of objects over time, and towards the end of the medieval period, the common understanding of time itself underwent a fundamental shift: a shift that enabled this new kind of science. All people are aware that certain activities or events have a characteristic duration, and life on this planet has always followed the beat of an ancient rhythm. Day after day the sun rises and sets, year after year the seasons are repeated, and it is surely obvious to everyone that a day is not the same as a year.

Ancient people could refer to periods of time (e.g. the time taken to boil a certain quantity of water), but it was the activities themselves that had a duration, and time was not typically conceived as an abstract concept, separate for the events whose duration we might care to measure. It is a characteristically modern sensibility that thinks of time as a fundamentally abstract sequence of units (days, hours, seconds, etc.). After all, at its root a clock is a mechanical representation of our planetary system: the clouds may cover the sun, but the hands of a clock will carry on ticking regardless, showing us the moment when the sun should be overhead. Only a modern person would ever think of saying that 24 hours is 24 hours, regardless of the rate at which the earth happens to revolve!

Clocks were first used in the monasteries of Europe, where the daily routine was highly regimented, and the desire to stick to a strict schedule of prayers was reason enough to keep careful track of time. Eventually the use of clocks spread from monasteries, to clock towers, and out into the cities beyond. This was a mixed blessing, as accurate time keeping enabled the rationing of time, and it permanently changed the way human affairs are organized. After all, clocks not only measure time; they are also used to synchronize the actions of men, telling us when to work, when to eat, and when the play begins.

In a world of clocks it seems self-evident that the passage of time is an empirical fact, and we can even come to believe that the facts that a clock can verify are somehow more real than our direct, subjective experience of the passage of time. In short, the psychological significance of the clock is that it separates time from human activity, or the events we can observe in Nature. As the writer and historian of technology, Lewis Mumford, remarked in *Technics and Civilization*, the spread of clocks 'dissociated time from human events and helped to create the belief in an independent world of mathematically measurable sequences: the special world of science.' This 'special world of science' is one in which individual objects of study can be measured and probed in isolation, and the relevant features for understanding are taken to be those that can be agreed upon by any suitably qualified panel of experts.

Objective measurement is central to the scientific enterprise, and in the words of the computer scientist Joseph Weizenbaum:

> This rejection of direct experience was to become one of the principal characteristics of modern science. It was imprinted on western European culture not only by the clock but also by the many prosthetic sensing instruments, especially those that reported on the phenomena they were set to measure by means of pointers whose positions were ultimately translated into numbers [e.g. barometers, thermometers or scales]. Gradually at first, then ever more rapidly and, it is fair to say, ever more compulsively, experiences of reality had to be represented as numbers in order to appear legitimate in the eyes of the common wisdom.

It was in this unquestionably measurable world that great thinkers such as Descartes, Newton and Leibniz made their mark. As we shall see in the following section, the youngest of these men, René Descartes, developed a philosophy and an approach to mathematics that were both radical and influential. This was a critical step in developing a science that went beyond that of the ancients, though as the preceding quotations should make clear, this move to modernity was not without a price.

Cartesian Coordinates

For the last four hundred years or so the study of geometry has become increasingly tied to algebraic methods, and it is this development that enabled Isaac Newton to describe the paths of moving objects in terms of an equation. This cross-fertilization between geometry and algebra was a massively important event, and not only because it gave rise to the use of equations in our descriptions of motion. In the words of the great mathematician Joseph Lagrange (1736–1813), 'As long as algebra and geometry travelled separate paths, their advance was slow and their applications limited. But when these two sciences joined company, they drew from each other fresh vitality and then forward marched at a rapid pace.' There are many mathematical, historical and philosophical factors behind this momentous shift, but François Viète (1540–1603) and René Descartes (1596–1650) are clearly pivotal figures.

As a man of undoubtable talent, Descartes was unusually hostile towards the other intellectual giants of his age. He was happy to work in relative isolation, and it seems that his preferred audience was one that would understand him without daring to challenge his sometimes erroneous views. For the most part Descartes' lifestyle was simple but leisurely, and up until the year of his death he refused to get out of bed before eleven o'clock. Unfortunately he agreed to teach Queen Christina of Sweden who wanted her philosophy lessons to begin at 5 a.m., and after a few weeks of early mornings in the depths of the Swedish winter, Descartes contracted a fatal case of pneumonia.

Like many great thinkers of his age, Descartes believed that we should rely on a 'general method of reason to find the truth in the sciences'. I would be tempted to dismiss this goal as absurd and unattainable, were it not for the fact that his attempts to summarize the general principles of valid reasoning revolutionized mathematical thought. One of Descartes' most influential innovations (which Pierre de Fermat independently conceived) was the idea of using coordinates, as when we draw a graph with an x- and a y-axis.

The story goes that as Descartes was lying in bed, he realized that the apparent location of the fly buzzing above his head could be described using two numbers: its distance from the wall behind his head, and its distance from the wall that stood to his left. This was a simple observation, but Descartes' genius was to recognize important ways in which this idea could be employed. Coordinate-based graphs are enormously convenient, as can be appreciated from their ubiquity in the modern world. Coordinates are also crucial for another, fundamental reason: they enable the practice of drawing a shape, while describing that shape through the use of a function (that is, a rule that takes a number as an input and gives a number as an output). In other words, coordinate systems make it possible to think of curves as functions, and to think of functions as curves.

For example, imagine drawing a graph of the function $y = x^2$. Once we have this graph, we can gain insights about the shape (a parabola) by studying and manipulating the equation. The Ancient Greeks had achieved the converse of this feat: they gained insights into equations by making geometric arguments. However, Ancient Greek 'equations' were not the concise, formal statements of modern mathematics. Instead, they wrote out relationships like $y=x^2$ using complete sentences. What is more, they considered such relationships to be properties of the underlying curves, and they did not consider the corresponding equations as mathematical objects in their own right. Consequently, they did not develop the basic algebraic techniques that Descartes had inherited from the Arabs.

For the Greeks, a curve was something that was traced by a moving point, and particular curves were produced by using a ruler and compass, by slicing a cone, or through some other idealization of a physical act. Following his idea of adopting a coordinate system, Descartes took the novel step of *defining* curves through his use of equations. Crucially, once we have defined a pair of curves $y=f(x)$ and $y=g(x)$, it is rather obvious that we can also consider the curve $y=f+g$, where the value of the function $f+g$ at any point x is simply $f(x)+g(x)$. Likewise, we can also consider the curves, $y=f-g$, $y=f\times g$ and $y=f\div g$, though the last of these curves will be undefined at any point where $g(x)=0$. Note that the addition, subtraction, multiplication and division of functions or curves are entirely new concepts. These important ideas are a natural extension of arithmetic using numbers, but the Ancient Greeks did not use symbols in this way, and they did not imagine that you could add or multiply curves.

Mathematics is now dominated by our use of discrete, algebraic symbols, but until the modern period, European mathematics was dominated by geometric reasoning. Descartes was an influential figure in bringing about this fundamental change, because of his philosophy as well as his contributions to mathematical technique. Because he was sceptical of the evidence of our senses, he thought that the real essence of mathematical statements lay in their 'rational' character. So, for example, if we were to discover that we are just brains floating in a jar, almost everything we thought we knew about the world would be wrong. Nevertheless, when we imagine this bizarre scenario we can also imagine doing maths in the same old way, and Descartes took this thought experiment as proof that the rule-governed use of symbols is our most certain form of knowledge.

Before Descartes' time, questions concerning linear equations were understood as being substantial precisely because those equations could be understood in terms of physical lengths. Similarly, quadratic terms could be understood in terms of areas, while cubic terms referred to volumes. We can still make those connections, but generally speaking, we don't. For example, we are happy to employ algebraic techniques to solve the equation '$x^2 + x = 110$'. Very few modern people would worry that x^2 represents an area, while x represents a length, and it doesn't make much sense to add an area to a length.

In contrast the Greeks, like the Egyptians and Babylonians before them, oriented their mathematics around geometric interpretations. For them, the equation '$a^2 + b^2 = c^2$' was invariably interpreted as a true description of the relationship between the areas of three different squares. Indeed, they didn't even write the equation that we know so well: they simply used words and diagrams to state the relationship between the relevant areas. Students today often think of Pythagoras' Theorem as primarily a recipe for generating a number given two other numbers. Of course, they know that the numbers in question refer to the lengths of a triangle, but in looking at the equation, they may not think of areas at all.

In short, Descartes made us recognize the fact that we can take a collection of lengths (or other measured quantities) as purely symbolic statements, divorced from any worldly sense of measurement. In other words, we can systematically name a length '2' instead of '2 cm' (say). The validity of this step marks the dividing line between mathematics and physics, because as far as abstract mathematics is concerned, the meaning of our units really isn't relevant. As a mathematician, Descartes was happy to consider adding x to x^2 or any other power, as we can remain oblivious to any intuitions that might motivate our use of any numerical symbol. For example, the expression x^2 does not have to be read as the area of a square of side x. It could just as easily apply to a line of length x^2, and as a person performing a given sum, we do not need to care which of these is 'really' the case.

This emphasis on the use of formal symbols is characteristically modern. Indeed, I would argue that for modern mathematicians it is the symbolic representability of concepts that seems to mark them out as grounded or 'properly defined', being parts of a symbolic system that others can employ. For example, the ancients had considerable conceptual difficulty in reconciling something like $\sqrt{2} = 1.414\ldots$ with their notion of number. I believe that there has been a subtle but significant shift in our sense of number, as most people today find it natural to think of a number as any sequence of digits whatsoever. In other words, symbolic forms such as '1.414...' are part of our system of numerals, and learning to use this system of numerals is at the heart of our initiation into the number concept.

At this point I want to clarify the history of coordinates and number lines, as I don't want you to get the impression that they were all invented in a single stroke. The use of coordinates, and the concept of longitude and latitude, had been used by astronomers and cartographers since Hipparchus (*c.* 150 BC). In fact, the idea of using coordinates while studying geometric curves dates back to Nicole Oresme (1323–1382), but Oresme did not develop this idea in quite the same radical fashion. What was new in Descartes' time was the use of axes *together with equations*, expressed in a concise, symbolic form. Descartes' understanding of the word 'line' was essentially classical, taking it to refer to the paths traced by moving points. His flash of insight was to realize that we can identify points in geometric space by their distance from the axes. The Greeks would have had difficulty developing this idea, as they did not have a concept of zero. They thought of numerical quantity in terms of counting and ratios between lengths, so they did not think of a line with a point corresponding to zero. It was Newton who added the innovation of having positive numbers on one side and negative numbers on the other (Descartes' graphs just had positive numbers), and as we shall see the modern concept of the number line was not fully formed until the end of the nineteenth century.

Linear Order and the Number Line

Number lines are a rather subtle concept, as it is far from obvious what numbers have to do with geometric lines. The most basic facts relate to a deep rooted, physically grounded understanding that applies whenever we have three objects, A, B and C, lined up in a row. We know that if object A is to the left of B, and B is to the left of C, it must also be the case that A is to the left of C. Similar statements also apply when we replace the words 'to the left of' with the words 'bigger than', 'higher than', 'hotter than' or 'heavier than'. We approach the world with the knowledge that we can put things in order of size, height, temperature or weight, and we understand that every such order is metaphorically akin to the spatial form of linear order.

This axiomatic truth concerning linear order is not some arbitrary convention like the rules of chess: the world seems to work that way, and it is very hard for us to imagine it being otherwise! Indeed, to use the number concept properly is to accept (among other things) that if x is smaller than y and y is smaller than z, then x must be smaller than z, just like the case of physical objects arranged in a line. That is to say, real numbers and points on a line both satisfy the following axioms:

1. Given a pair of numbers x and y, either $x<y$, or $x>y$, or $x=y$. Likewise, if we are given a couple of points on a line, either the first point is before the second, or the first point is after the second, or both points are in fact the same point.
2. If $x<y$ and $y<z$, then $x<z$. Likewise, if point x is to the left of y and y is to the left of z, then x is to the left of z.

In short, the relationships between numbers are structured in the same way as the relationships between points on a line. On the other hand, numbers and the points on a line are fundamentally different because geometric lines are continuous, with no gaps between their parts, while numbers are essentially discrete. Although the number line is 'continuous' in the sense that it is not missing any real numbers (i.e. the number line does not have a gap located at any real number), the number line is not continuous in the geometric sense, because the parts of a number line are points, and these parts do not touch. Points on a number line cannot touch, because if there is no distance between a pair of real numbers (as there would be if they touch), they are in fact one and the same number/point.

Crucially, once we have identified one point as 'being' the number 0 and another point as 'being' the number 1, every point on a line determines a unique real number. More specifically, there is some distance between the point in question and the point 0, and there is some other distance between the points 0 and 1. Given any point, the ratio between those two lengths determines a quantity in the classical sense, and we can think of that number as being the distance between the point and the origin (that is, the point that we identify with 0). Conversely, we can use the concept of 'distance from the origin' to associate each real number with some or other point on the line, with different real numbers picking out different points in space. It was Richard Dedekind who explicitly made the final conceptual leap, when he asserted that *we can define the number line as being the set of real numbers*, where the point identified by the number r is r units from the origin.

This new definition says that fundamentally, a line is an infinite set of points. That is subtly but significantly different from the classical, geometric conception of a line, which says that a line is the path traced by a moving point. In the case of the classical, geometric line, the lines themselves are primary, and points on a given line are merely infinitely precise locations on that line, not the thing that the line is made of. Furthermore, in classical geometry the lines themselves are understood to have certain intrinsic properties. For example, a line that forms a closed loop constitutes the boundary of a region in space, and we understand that this is so because of our conception of lines, rather than the points that are on that line.

By Dedekind's definition the points are always the primary entities. Lines are redefined as being infinite sets of discrete points, and in this framework the properties of lines need to be redefined in terms of relationships between those points. In fact, because points have zero length, there is no end to the number of points that we can fit onto a geometric line, and we are not obliged to assume that there is one and only one point for every real number. However, by adopting Dedekind's definition we acquire a system in which the arithmetic of points on a line is fully determined, and, what is more, we can prove all of classical geometry within this new, fundamentally discrete system.

Isaac Newton

Newton's father was an illiterate free-hold farmer who died months before his birth. When his mother Hannah Ayscough remarried some three years later, Newton's step-father insisted that she leave her son with the Ayscough family. Newton grew up to be a rather neurotic and secretive individual, prone to fits of rage. He never married, loathed criticism, and was intensely curious about the Bible and the world around him. Indeed, his curiosity about the nature of man's vision almost drove him blind. Among other things, he carefully described what happens when you stare at the sun all day, and how your field of vision distorts when you squeeze your eyeballs with a wooden stick. When the plague struck Cambridge in 1665, a young and largely unnoticed Isaac Newton (1642–1727) left his lodgings at Trinity College and moved back to Woolsthorpe, the Lincolnshire village of his birth. Although Newton was only in his early twenties, he at his creative peak, there is a strong case for arguing that no scientist ever achieved more in a two year period than Newton did between 1664 and 1666. In this brief period Newton conceived of three of his greatest insights (any one of which would have earned him a prominent place in history): the Law of Gravity, the idea that white light is physically comprised of all the colours of the rainbow, and the beginnings of the calculus.

Unlike his more speculative predecessors, Newton did not try to explain *why* things moved, he simply focused on giving a mathematical account of *how* things move, in the language of force, mass and acceleration. While Aristotle and his followers had tried to distinguish between heavenly motion and the motion of things on earth, Newton boldly stated that all matter attracts all other matter, and this 'gravitational' attraction is a truly universal force. That is to say, what Newton realized in a flash of insight was that the sun and the earth are not the only things that exert a gravitational pull, as an apple or a lump of soil or rock attract mass just like any other object (of the given mass). Furthermore, given any pair of objects, the size of the force of gravity will be proportional to each of the masses, and inversely proportional to the square of the distance between them. In other words, where the first object has mass m, the second object has mass n, and they are a distance r units apart, the force F pulling the two objects together will be

$$F = G\frac{mn}{r^2}$$

where G is a number known as the 'universal gravitation constant'. This simple statement is almost a complete summary of Newton's famous theory. All we are missing is a definition of what happens when an object experiences a force of a given size. The question 'What is a force?' is far from trivial. Nevertheless, to understand the success of Newton's theory, we only need to understand the motion that forces produce. By definition, if an object of mass m is subject to a force of size F, that object will accelerate at a rate F/m.

Newton's Law of Gravity has been justifiably described as the greatest generalization achieved by the human mind. This simple 'inverse square law' has many implications, the first of which is easy to deduce. Suppose that we have two objects of different mass, located in the same place on earth. Newton's Law implies the result that Galileo first suggested: both objects will accelerate at the same rate, namely Gm/r^2, where m is the mass of the earth, and r is the distance from the objects to the centre of the earth.

The very same law can also be used to explain the motion of the planets, the path taken by a projectile, the existence of the tides, the eccentric orbits of comets, the precession of the earth's axis, and many other measurable phenomena. In particular, the German astronomer and mathematician Johann Kepler (1571–1630) made a famous study of some very detailed observations of the motion of the planets across the night sky. He tried to summarize this information in a number of ways, and in 1609 he finally formulated a simple scheme that fitted his data extremely well. Kepler's three famous rules are as follows:

1. As the planets move about the sun, they make the shape of an ellipse, with the sun at one of the foci.
2. If you draw a line from a planet to the sun, and measure the area that is swept out in a fixed period of time, you always get the same answer.
3. If you square the amount of time it takes for a planet to complete an orbit, and divide this number by the cube of the width of the planet's ellipse, you always get the same number (a constant that depends on the weight of the sun). Among other things, this means that the length of a planet's year can be calculated from its average distance from the sun, and vice versa.

Famously, all of these results are logical consequences of Newton's Law of Gravity. Indeed, it should be emphasized that Newton's singular achievement was not so much conceiving the Law of Gravity, as elucidating its mathematical consequences. Indeed, we know that in 1684 Edmond Halley, Sir Christopher Wren and Robert Hooke met in a London coffee shop to discuss the idea that the motion of planets is governed by an inverse square law, and at that time Newton's ideas could be found only in his private notebooks. Halley, Wren and Hooke tried and failed to calculate the path that a planet will follow if it is subject to an inverse square law, and later that same year Halley asked Newton if he knew how to solve the problem. Newton immediately responded that an inverse square law produces elliptical motion, as he had known that result for nearly twenty years. Halley was delighted by Newton's response, but when he asked Newton for a rigorous proof, it took him nearly three months to reconstruct this argument from his youth. Halley recognized the supreme importance of Newton's work, and thanks to his encouragement and financial backing, Newton was persuaded to spend the next year and half writing the most influential physics book ever written: the *Principia Mathematica.*

Newton could not have written his masterpiece without developing calculus. In light of this fact, it is somewhat surprising to learn that *Principia* does not directly employ the language of calculus. Instead, Newton relied on geometric arguments that were very much in the tradition of Ancient Greece, combined with physical intuitions about motion, and appeals to geometric notions of a limit case. That is to say, as far as possible, Newton used very traditional means for finding tangent lines or the areas under curves. Nevertheless, to develop the new science of mechanics, Newton needed to answer questions that were beyond the Ancient Greeks. In particular, he needed a way of handling the notion of a rate of change. For example, it is clear that if an object moves four metres in one second, it is moving with a speed of four metres per second. Similarly, if an object moves two metres in half a second, it is moving with a speed of four metres per second. But how can we define the speed of an object *in an instant*, when it doesn't have time to move any distance at all?

The Fundamental Theorem of Calculus

As far back as the fourteenth century, the French mathematician Nicole Oresme realized that we can represent the motion of an object with changing position and velocity through the use of a graph. For example, we can use the x-axis to represent time, while the y-axis represents distance from the origin. As we move from left to right, our curve can go up or down, representing the changing location of the object in question. The faster our object moves, the steeper our curve will be. Indeed, given such a representation of motion, we can *define* the instantaneous speed of our object to be the gradient of this curve. This idea connects mechanics with one of the two fundamental procedures of the calculus: finding the gradient of a curve (differentiation). The other fundamental procedure of the calculus is called integration. In many cases we can identify the process of integration with the process of finding the area under a curve, though strictly speaking that is not how integrals are defined.

Archimedes and other ancient mathematicians were familiar with the concept of a tangent. They also knew how to find the areas underneath certain curves. However, their proofs were quite specific to the shape in question, and you couldn't easily adapt their arguments when confronted with a new curve. Newton and Gottfried Wilhelm Leibniz (1646–1716) independently developed calculus, which is a much more general method. Over the centuries calculus has been used in countless ways, but the simplest thing we can use it for is to calculate the areas and gradients of a curve, given an equation that describes that curve.

As a simple example, imagine that we wish to find the gradient at a point on a parabola. Because our graph is perfectly described by the equation $f(x) = x^2$, we can find the gradient by examining the equation. This is in contrast to the kinds of geometric reasoning that would have occurred to the Greeks. The graph becomes increasingly steep, so we can find a line that will be slightly steeper than the tangent by drawing a line between the points (1, 1) and $(1+d,(1+d)^2)$. Some simple algebra shows that this line has a gradient of $2+d$ (for any number d greater than zero), which means that the tangent line must be smaller than any number of this form.

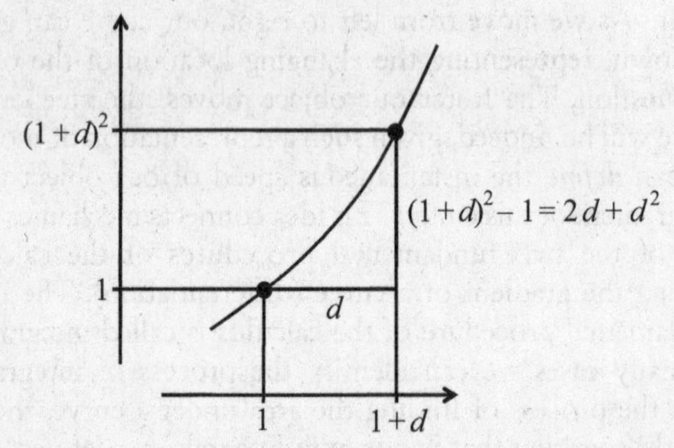

To draw a line between these two points, we move $2d+d^2$ units up for every d units that we move across. Therefore the gradient of the line between these two points is:

$$\frac{2d+d^2}{d}=2+d.$$

Similarly, the tangent line must have a gradient greater than $2-d$, for any number d greater than zero. Therefore, the tangent line has a gradient of 2 (no more, no less). More generally, the tangent line at the point (x, x^2) has a gradient less than $2x+d$ but greater than $2x-d$, which implies that the gradient is equal to $2x$.

In other words, the gradient at the point (x, x^2) is $2x$. We now have a second equation, namely $f'(x)=2x$, which tells us the gradient at any point on our original curve

$f(x)=x^2$. The equation f' is called the derivative of f, and as we shall soon see, the equation f is called the integral of f'. Note that we have found f' by analyzing f: making a deduction based on the way that we multiply out $x+d^2$. There is also a second connection between the equations $f(x)=x^2$ and $f'(x)=2x$:

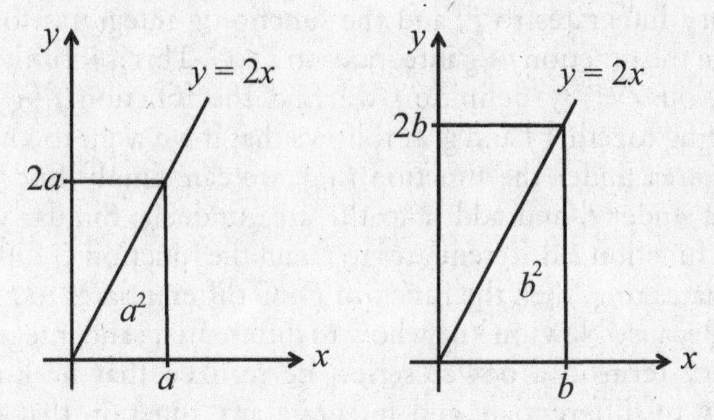

The area between the x-axis, the y-axis, the line $y=2x$ and the line $x=a$ is equal to a^2 for all values of a. In the notation of Leibniz, we write:

$$\int_0^a 2x\,dx = a^2.$$

In 1684, Leibniz published the Fundamental Theorem of Calculus. In other words, he proved that finding the area under a graph (integration) is the converse of finding the gradient of a graph (differentiation). We will examine Leibniz's theorem below. First I want to mention that Newton had been familiar with this result for about eighteen years. Newton was able to make this step because he had immersed himself in the new mathematics of algebraic geometry. In particular, he was a virtuoso at working with 'power series'. That is to say, he learned that certain geometric and trigonometric relationships were expressible in terms of sequences of non-negative powers of x. For example, like the mathematician-astronomers of Medieval Kerala, Newton knew that:

$$\sin(x) = x - \frac{x^3}{3 \times 2 \times 1} + \frac{x^5}{5 \times 4 \times 3 \times 2 \times 1} - \ldots$$

Newton also knew another fundamental fact: if the function f integrates to F, and the function g integrates to G, then the function $f+g$ integrates to $F+G$. This is intuitively obvious, as by definition we find the function $f+g$ by adding together f and g. It follows that if we want to know the area under the function $f+g$, we can simply find the area under f, and add it to the area under g. Similarly, if the function F differentiates to f, and the function G differentiates to g, then the function $F+G$ differentiates to $f+g$.

Because Newton knew how to differentiate and integrate every term in a power series, he realized that he knew how to differentiate and integrate any function that can be expressed as a power series. At least, that is how I would describe his approach: Newton himself did not use the words function, integrate or differentiate. Nevertheless, in 1666 Newton prepared a manuscript entitled *On Analysis by Equations with an Infinite Number of Terms*, and he showed it to a couple of his peers. Several years later Newton was ready to publish his ideas on calculus as a technical appendix to a book concerning optics, but following a dispute with his hated rival Robert Hooke, Newton withdrew the entire work.

Newton was obsessed with the ideas that gripped his mind, and he alternated between being suspicious, indifferent or disdainful of what the wider world might think. In contrast, the great polymath Leibniz always had one eye open for an appropriate audience, and he would happily work on any problem that might earn him money or enhance his reputation. Like Newton, Leibniz found a way to use the principles of arithmetic in thinking about rates of change, but there were significant differences in the forms of calculus that the two men developed. Newton argued geometrically, looking at the limit case of a sequence of tangents. In contrast, Leibniz's argument explicitly referred to a ratio of infinitely small quantities.

From a modern perspective, the concept of an infinitely small quantity is perfectly respectable, even though (by definition) such infinitesimals are smaller than any real number. I say that infinitesimals are perfectly respectable because they have a clear definition in set theoretic terms, yielding a number system called the 'hyperreals', and a form of mathematics called 'non-standard analysis'. However, as that name suggests, mathematicians do not usually admit infinitesimals as part of their number system. By using the concept of a limit case we can do calculus without referring to infinitely small 'numbers', and since that is the standard method, modern mathematicians follow Newton rather than Leibniz. On the other hand, it is Leibniz's superior notation that we use today.

Leibniz's proof of the Fundamental Theorem of Calculus is particularly elegant, and it exemplifies his remarkable ability to spot the crucial point. We begin by imagining a continuous, unbroken curve described by an equation $f(x)$. Any mathematical rule that takes a number as input and gives a number as output is called a 'function', and our equation $f(x)$ is a function because for each point a there is unique output, namely $f(a)$. As shown below, we can turn this equation into a drawing of a curve, where the height of the curve at the point a is $f(a)$. Note that at each point, there is a certain amount of distance between the x-axis and our curve. Furthermore, we can enclose a finite area between the x-axis, the y-axis, the curve $y=f(x)$ and the line $x=a$. Leibniz denoted this area with the symbols $\int_0^a f(x)dx$, but for simplicity's sake, I will use the symbols $F(a)$ to denote the area between 0 and a.

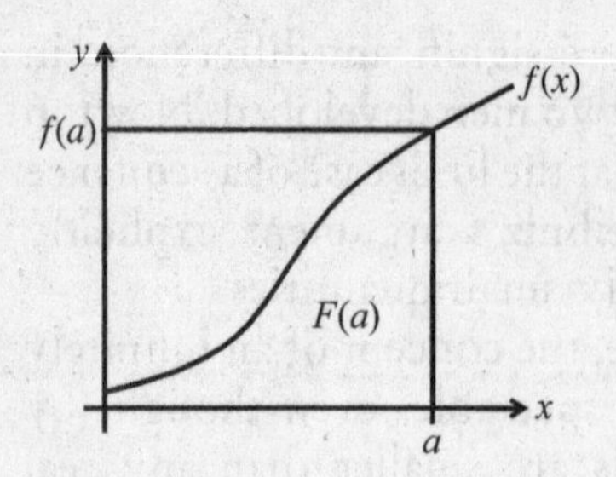

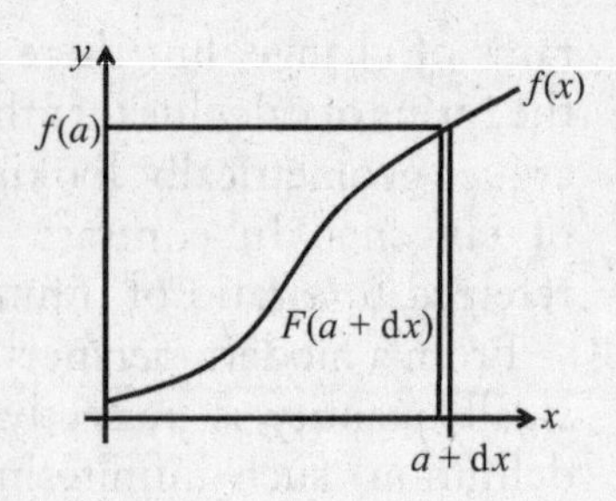

If we vary the constant *a*, we change the area in question. In other words, our original curve also determines a second function, as for any positive number *a* there is a unique number F(*a*), which is equal to the area under the curve up to the point *a*. Now, if we change *a* to $a+dx$ (where dx is an 'infinitesimal' amount), the area in question gets infinitesimally larger. Assuming that $f(a)$ is positive, $F(a+dx)$ is larger than F(*a*), because the area $F(a+dx)$ contains an additional strip of height $f(a)$ and width dx.

By definition, the gradient of $F(x)$ at the point $x=a$ is given by a ratio: the change in $F(x)$ as we increase x from a to $a+dx$, divided by the change in x (a change that equals dx by definition). In other words, the gradient of $F(x)$ at the point $x=a$ is $\frac{F(a+dx)-F(a)}{dx}$. We have just seen that $F(a+dx)-F(a)=f(a)dx$. Hence the gradient of $F(x)$ at $x=a$ is:

$$\frac{F(a+dx)-F(a)}{dx}=\frac{f(a)dx}{dx}=f(a).$$

Since $F(x)$ is the integral of our arbitrary, continuous function $f(x)$, it follows that if we integrate a function and then differentiate the result, we get back to our original function. In other words, the Fundamental Theorem of Calculus holds true.

From Algebra to Rates of Change

In the previous section, I recounted Leibniz's proof of the Fundamental Theorem of Calculus. Apart from being in English, my version of the argument is very similar to his original proof. There is, however, a subtle but significant difference in the way that we follow the argument. Unlike seventeenth-century mathematicians, modern readers will be familiar with the concept of a function. Strictly speaking a function is a kind of set, where each acceptable input is paired with a unique output. Even if we are not familiar with this formal definition, we are used to the idea of rules that map inputs to outputs. For example, when we learn how to square numbers, we might be told that $f=x^2$ 'is a function'. Consequently, we now think of functions as basic mathematical *objects*. For earlier generations, finding the area of a square of side x was a *process*, while the objects were the numbers (or areas) themselves.

Leibniz was an important figure in bringing about this change. As a philosopher, he was fascinated by the idea that we can capture knowledge in the form of a rule, and I think it is fair to say that in some sense, Leibniz had a function-based perspective. Nevertheless, the modern, totally general definition of a function was not put forward until 1755, when Leonhard Euler wrote his *Institutiones Calculi Differentialis*. Indeed, one could argue that the concept was not fully developed until the work of David Hilbert (1862–1943). In particular, Hilbert wrote about 'operators', which are rules that take a function as an input, and give a function as an output. From his modern, highly abstract perspective, the set of all real-valued functions is a typical mathematical object, much like the set of all real numbers.

Integration and differentiation are both operators, as when we differentiate a function, we produce another function. For example, we have seen that the function $f(x)=x^2$ differentiates to $f'(x)=2x$, and that $f'(x)=2x$ integrates to $f(x)=x^2$. It is also worth mentioning that other basic calculus results were known to mathematicians before Newton and Leibniz. For example, the Dutch mathematician Johann van Waveren Hudde (1628–1704) knew that if we differentiate the function x^k, we get kx^{k-1}. Of course, he didn't use the term 'differentiate', but he calculated the tangents of some algebraic, geometric curves, including ones of the form $y=x^k$. As I have mentioned, this kind of equation-based approach to geometry was the form of mathematics that immediately preceded the invention of calculus, and Leibniz and Newton were both familiar with Hudde's work.

An even older result is that the area under a curve x^k is equal to $\frac{x^{k+1}}{k+1}$ (provided that k is not equal to -1). We can deduce this fact by combining the previous result and the Fundamental Theorem of Calculus. Also note that particular cases of this result correspond to standard, geometric problems. For example, finding the area under x^2 corresponds to the problem of finding the area under a parabola. Indeed, the special cases where $k=1$, 2, 3 or 4 were known to the Arab mathematician Ibn al-Haytham (*c.* 965–1039). Six centuries later, the Italian Bonaventura Cavalieri reinvented Archimedes' idea of slicing a shape into infinitely many pieces, and proved this result for values of k up to 9. He conjectured that it is true for all integers k, and in the 1630s the French mathematicians Fermat, Descartes and Roberval proved that his conjecture was correct. Before calculus was even invented, Fermat had effectively proven that $\int_0^a x^k dx = \frac{a^{k+1}}{k+1}$ for fractional values of k.

In the centuries that followed the invention of calculus, European science underwent a massive expansion. As Newton had forcefully demonstrated, calculus and the new mathematics could prove invaluable to science. As ever, the developing mathematical theory co-evolved with the questions people sought to answer. In particular, later generations of mathematicians developed the theory of differential equations. By definition, a differential equation is one that relates a function f to its derivative f'. For example, $f(x) = \frac{x}{2} f'(x)$ is a differential equation (though not a very interesting one), and we can see that $f(x) = x^2$ is a solution. Differential equations play a prominent role in engineering, physics, economics, chemistry, biology and many other disciplines. In some ways this is not surprising. Whenever we have a rule that tells us how a quickly continuous quantity changes over time, we almost always end up producing a differential equation.

For example, if we have a few bacteria in a dish, it is reasonable to assume that the number of bacteria 'born' per minute will be proportional to the number of bacteria that are currently in the dish. In other words, if there is no shortage of resources, it is reasonable to model the population size by saying that the derivative f' is proportional to f, where $f(t)$ is the population size at time t. This is one of the simplest differential equations, and if the constant of proportionality is a positive number, the solution to this equation is exponential growth. In other words, our assumption about the rate of change f' implies that if the number of bacteria has increased from n to $2n$ by time t, there will be $4n$ bacteria at time $2t$, $8n$ at time $3t$, $16n$ at time $4t$, and so on.

A more interesting differential equation is described when we specify that taking the derivation of the derivative of our function f produces an equation that is proportional to our original equation f. This kind of differential equation can be used to describe a plucked string, as when we pluck a guitar string, it accelerates back towards its starting point at a rate that is proportional to the distance by which the string has moved. So, for example, parts of the string that have moved a distance $2x$ from their starting point accelerate towards their starting point at twice the rate of those parts that have been moved a distance x. This particular differential equation is known as the 'Wave Equation', and it is one of the most important equations in the physical sciences. The initial questions that inspired this equation came from the study of vibrating strings, as people wanted to describe and explain all the different ways that a string could be made to vibrate. More specifically, finding 'a solution to the Wave Equation' meant finding one possible form of vibration, while 'solving the Wave Equation' meant finding every possible solution.

Several mathematicians tried to summarize all the possible forms of vibration, and arguments over the best approach helped to extend and clarify the concept of a function. For example, if we look at my account of Leibniz's proof of the Fundamental Theorem of Calculus, it isn't really clear how free we can be in picking our function f. The argument certainly holds true for polynomial equations, but can we imagine that f is any rule for pairing an input with an output? To put it another way, just how arbitrary is an arbitrary, continuous function? Similarly, until the language of analysis was refined, it wasn't clear whether every 'function' could be expressed in terms of a power series.

We now know that the Wave Equation has all manner of applications. It plays a crucial role in describing and predicting the behaviours of liquids, gasses, electromagnetic phenomena and many other things. Indeed, it may be fair to say that vibrating strings have inspired more mathematics than any other object. Many different branches of maths, from partial differential equations, to Fourier series and set theory, have deep roots in the study of vibrating strings. It is also remarkable to note that with our current knowledge, it has become clearer than ever that if you want to study the general phenomenon of vibration and oscillatory form, stringed instruments are the best place to start (much easier, say, than studying the vibrations of the skin of a drum).

The fact that mathematicians have been studying stringed instruments for thousands of years takes on particular significance when we reflect on the supreme importance of selecting the objects of study that yield the mathematical models of the greatest simplicity. After all, finding a particular, easily managed case, analyzing it, and using that analysis to develop a broadly applicable framework or theorem is a central and recurring theme in all of our attempts to understand the world. That is because when we are faced with a problem that we do not know how to solve, we usually make progress by recognizing that the problem at hand is somehow akin to some other problem that we do know how to solve. Because of this pervasive pattern, relatively simple, exemplary cases are an essential part of every kind of expertise (including mathematics), as our general sense of how the world works is deeply grounded by our understanding of particular, concrete cases.

The importance of simple, concrete cases is sometimes underplayed in the case of mathematics, as we might reasonably claim that if mathematics is 'about' anything, it is about general principles rather than particular, concrete cases. However, that does not mean that special cases are anything less than essential. We might always want to move on to the most general statement we can make, but as Hilbert remarked, 'The art of mathematics consists in finding that special case which contains all the germs of generality.'

Chapter 6:
LEONHARD EULER AND THE BRIDGES OF KÖNIGSBERG

> 'True eloquence consists in saying all that is required and only what is required.'
>
> François de La Rochefoucauld, 1613–1680

Leonhard Euler

Leonhard Euler (1707–1783) is widely recognized as being one of the greatest and most prolific mathematicians to ever live. He is famous for many reasons, and has often been called the father of modern mathematics. The mathematical notation that we use today is largely due to Euler (pronounced 'oiler'), and because Euler was one of the first people to understand the power of the calculus, he was able to make many major contributions to physics, engineering and astronomy.

During Euler's lifetime, the *Grand Prix* of the Paris Academy of Sciences was the most prestigious and lucrative award a scientist could win. When he was only nineteen, the Paris Academy posed the following question: 'What is the best way to arrange the sails on a ship?' Euler was born in Basel, and had never left the land-locked nation of Switzerland. Despite only having seen pictures of large, sailing ships, the young Euler conceived of the concept of 'an equivalent sail'. Essentially, this is the position that a single mast and sail would need to be in if they were to match the net propulsive force of some collection of masts and sails.

At the time, the mathematical study of forces and Newton's laws of motion were novel innovations, but Euler was supremely confident in the new science of mechanics. His 'Thoughts on a Nautical Problem' would surely have shocked many an experienced shipbuilder, as the teenager boldly stated that 'I did not find it necessary to confirm this theory of mine by experiment because it is derived from the surest and most secure principles of mechanics, so that no doubt whatsoever can be raised on whether or not it be true and takes place in practice.' Sure enough, Euler's analysis was sound, and his ideas were incorporated into the next generation of English and French naval ships.

Euler continued to publish a stream of brilliant ideas right up until the day he died, even though he was completely blind for the last twelve years of his life. He was uniquely willing to share his preliminary guesses and partial proofs with anyone who was interested, and no mathematician in history has been more prolific.

Unlike many mathematicians, he constantly sought to explain himself as simply and directly as possible, often working to prove and prove again already established results. He wrote the most influential textbooks since Euclid, and did more to create our efficient, modern notation than any other author. For example, his particular way of writing the facts of trigonometry is that taught today: he started the practice of writing $f(x)$ to denote a function of a variable x, and he introduced the symbols e for 2.718... and i for $\sqrt{-1}$.

In Euler's time it was just about possible for one individual to learn the entire body of European mathematical knowledge. This is what Euler did, and he possessed an exceptional ability to cut through the clutter of centuries' of mathematical practice, revealing the necessary concepts through particularly perspicuous representations. Indeed, I think it is fair to characterize his genius by saying that more than any mathematician that preceded him, Euler lived by the principle that Wittgenstein articulated: 'The mathematician has no right to be surprised.' When experimental scientists put nature to the test, they must keep their eyes open for surprising results. By way of contrast, mathematicians are struck by the fact that established concepts may be far more fruitful and rich in patterns than anyone had realized.

For example, in the case of Pythagoras' Theorem we may start by being surprised by the regularity in the ratio of sides, because we don't yet understand the result. However, the effect of the proof is to make us see the result as obvious, as we might say: 'It's no surprise that the area of a^2 plus b^2 equals c^2, since all you have done is rearrange the triangles.' Many of the greatest mathematicians have used their understanding to penetrate a mystery, and presuming themselves exceptionally able, they have left the result as something difficult to see. Euler's genius was to realize that if mathematicians are convinced of a particular conclusion, it is wise to have faith that the relevant reasoning can be *stated* with a simple precision. Time and time again he searched for more ever direct ways of proving already established results, and he is credited with crisply formalizing much of the mathematics and physics of his day.

The Bridges of Königsberg

Königsberg was a town in Prussia, but it is now an enclave of the Russian Federation, and has been renamed Kaliningrad. The old town of Königsberg was famous for its seven bridges, and in the early eighteenth century, people sometimes posed a challenge: 'Can you walk through town crossing every bridge once and only once?' Euler's analysis of the problem is a beautiful example of a mathematician ignoring the irrelevant, saying all that is required and only what is required. More specifically, he realized that it doesn't matter how you wander about within each of the districts in town. We can draw each bridge as a line and each district as a point, and the resulting map captures all the relevant facts. In other words, we can move about the town crossing each bridge once if and only if we can move about the following simplified map by crossing each line once:

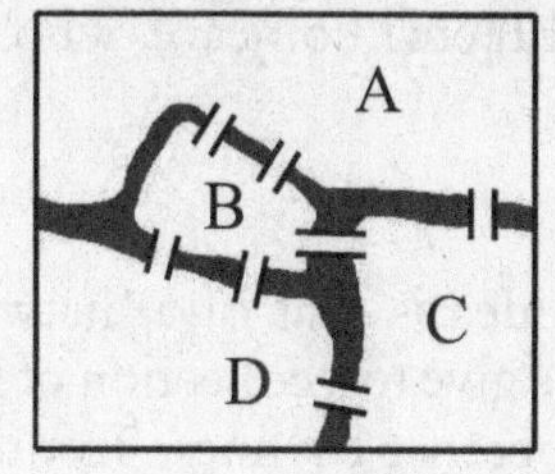

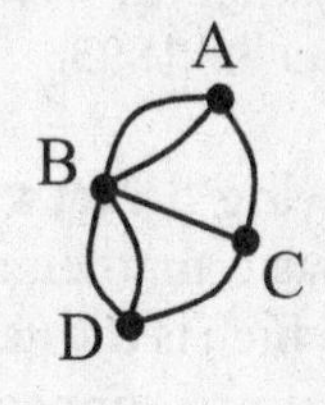

We have to start in a district, and we have to stop in a district, and there are four districts in Königsberg. This tells us that there are at least two districts where we neither start nor finish. Imagine someone trying to find a path that crosses each bridge once. As they visit one of the districts where they neither start nor finish, they must leave on a different bridge to the one they used to enter, so each 'entry' bridge must be paired with a different 'exit' bridge. This tells us that any region where we neither start nor finish must have an even number of bridges. In other words, to complete the task of crossing each bridge once, the districts where we neither start nor finish must contain an even number of bridges. All four regions in Königsberg have an odd number of bridges, which means that the task must be impossible.

As well as being a charming little puzzle in its own right, Euler's solution pointed towards some new and beautiful maths. Indeed, when Euler published his thoughts about the problem, he entitled his paper, 'The Solution of a Problem Relating to the Geometry of Position'. This is a very suggestive title, and in the next chapter we will see how mathematicians developed the idea of multiple geometries. First, I want to explore the concept of a network, and the notion that there is a subset of geometric truths, namely the truths where distance is not relevant. Euler had some understanding of this powerful idea, but as we shall see in a later section, a rigorous and fully general presentation of the basis of 'topology' had to wait until Henri Poincaré, who tackled the subject in 1895.

On Drawing a Network

A very basic mathematical idea is that of a 'network' or 'graph', which is the name we give to a collection of labelled points that are connected together by lines. To qualify as a single, connected network, we must be able to reach every point by crossing the lines of the network. Furthermore, the end of every line counts as a point, as does every intersect point (where two or more lines meet). To develop this idea further, we need to notice that we can draw any network (starting with a point) using three different types of move:

1. Draw a new line to a new point.
2. Draw a new point on an old line.
3. Draw a new line between two old points.

Now, imagine any connected network. A simple way to characterize that network is to count the number of specified points (giving a number P), and count the number of lines L. We can also count the number of regions R that a drawing contains. Intuitively speaking, a region is just an area contained within a set of lines. So, for example, if our network was a square with a point at every corner, we would have P = 4, L = 4 and R = 1.

We now have three numbers which we can use to describe any given network. To gain a further insight, let's look back at our three network-generating rules. After move (I), we have one extra point and one extra line, so our total number of points P goes up to P + 1, while the total number of lines L goes up to L + 1. The same is true of move (II). As you can verify with a pencil and a piece of paper, move (III) always gives us one extra line and one extra region, so L goes to L + 1 and R goes to R + 1. We can now make a startling observation: after any of our three moves, the number P − L + R remains the same. In other words, no matter what we draw on our piece of paper, P − L + R must always equal one!

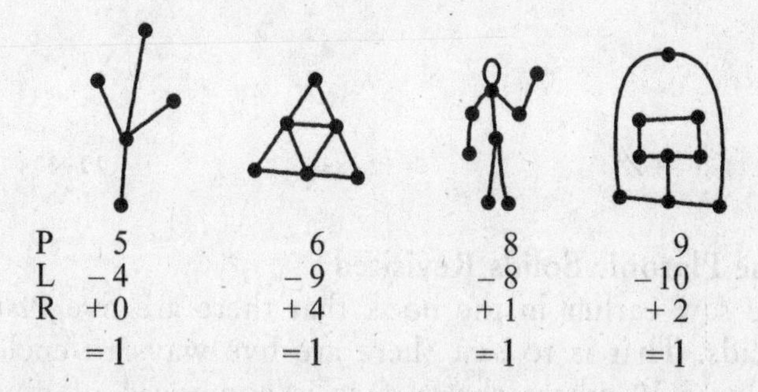

P	5	6	8	9
L	−4	−9	−8	−10
R	+0	+4	+1	+2
	=1	=1	=1	=1

All of the networks we have considered have two things in common. First, they have an 'Euler number' P − L + R = 1, and, second, they are all drawn on a piece of paper. Suppose that we had drawn them on a sphere instead. In that case we would start our drawing with one region and one point, as opposed to no regions and one point. This tells us that the Euler number of a drawing on a sphere is 2. This is true of every drawing on a sphere. The argument is almost exactly the same as for networks on a piece of paper, it is just that the drawing we start with (a single point) has an Euler number of 2 instead of

1. In particular, we can draw the edges of a cube onto a sphere: just imagine a cubic balloon that has been over-inflated into a sphere. The important point is that a cube that has been inflated into a sphere has the same number of faces, corners and edges as any other cube. Our theorem also applies to the other Platonic solids. These kinds of shape have their own vocabulary, so instead of counting 'numbers of points' we talk about 'number of corners', 'number of lines' becomes 'number of edges', and 'number of regions' becomes 'number of faces'. Our analysis of drawing networks shows that all manner of faceted solids have an Euler number of 2.

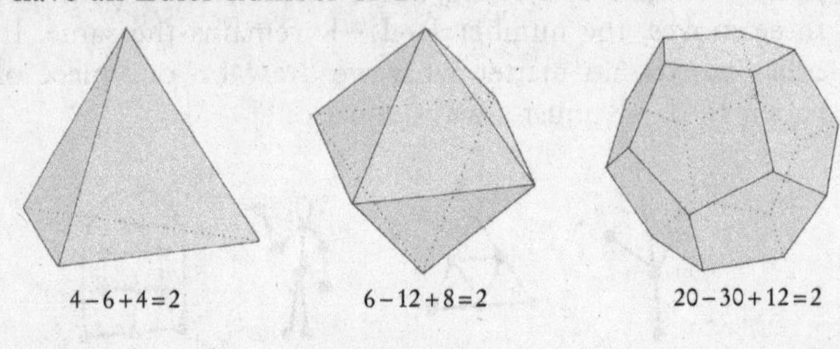

The Platonic Solids Revisited

We saw earlier in the book that there are five Platonic solids. That is to say, there are five ways to enclose a volume V, where the surface is composed of multiple copies of a regular polygon P. At each vertex the total angle of the polygons must add up to less than 360°, and there are only five cases that can satisfy this demand. However, we have not yet proved that there is only one regular shape for each of the following types of corner arrangement:

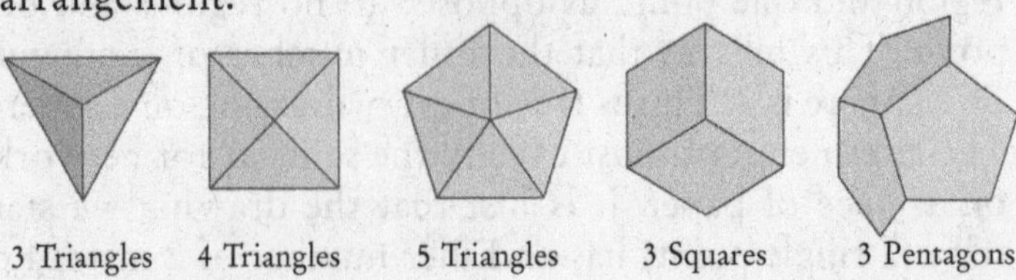

The completed shapes are described by three numbers: the number of corners, the number of edges, and the number of faces. There are three more numbers that are relevant: the total number of polygons (denoted R), the number of polygons that meet at each corner (denoted N) and the number of corners or sides on each of the polygons (denoted S). If we scatter the pieces on the ground, we have a total of RS edges and RS corners (R polygons, each of which has S edges). On the completed shape, every edge is shared by two different polygons. This means that the total number of edges must be $RS/2$. Furthermore, each corner is shared by N different polygons, so the number of corners on our completed shape is equal to RS (the total number of corners on the polygons) divided by N (the number of polygons is takes to make one corner of the finished shape). We now have three equations, which tell us the number of corners or points (P), edges or lines (L) and faces or regions (R):

$$L = \frac{RS}{2}, \ P = \frac{RS}{N} \ \text{and} \ R = 2 + L - P = 2 + \frac{RS}{2} - \frac{RS}{N}.$$

There is only one possibility for the number of faces, because the number of faces (R) is uniquely determined by the numbers S and N. For example, using pentagons at the corners we have $S=5$ and $N=3$, so we know that the number of faces on the completed shape satisfies the following conditions:

$$R = 2 + \frac{5}{2}R - \frac{5}{3}R \ \text{and this implies that} \ R = 12.$$

Since we now know the value of R, our equations also tell us that $P = \frac{5}{3}12 = 20$ and $L = \frac{5}{2}12 = 30$.

Given that we can construct the five Platonic solids, we can see that there are precisely five solids with surfaces consisting of a number of identical polygons, meeting at identical corners.

	Number of Faces	Number of Corners	Number of Edges
Tetrahedron	4	4	6
Cube	6	8	12
Octahedron	8	6	12
Dodecahedron	12	20	30
Icosahedron	20	12	30

Notice that cubes and octahedrons both have twelve edges, but one has six faces and eight corners, while the other has eight faces and six corners. This observation is closely related to a beautiful fact: if we draw lines connecting the centre points of each of the faces of a cube, we generate an octahedron, and if we connect the centres of an octahedron, we generate a cube:

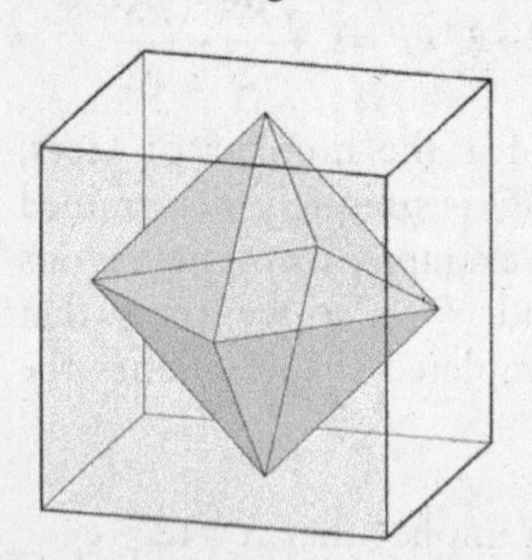

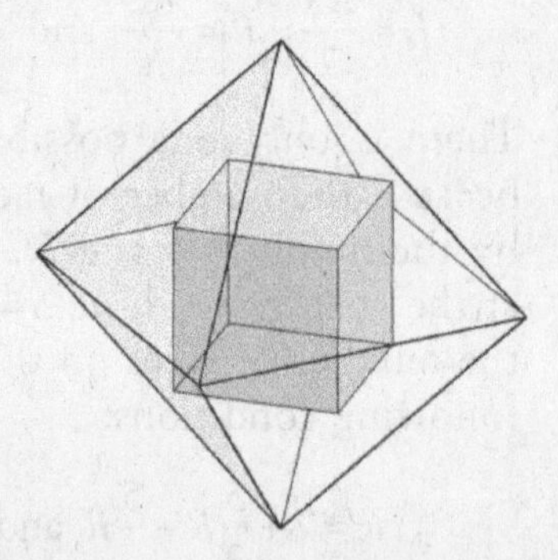

Because of this intimate relationship, we say that these shapes are dual. As the cube is dual to the octahedron, so the dodecahedron is dual to the icosahedron, while the tetrahedron is dual to itself:

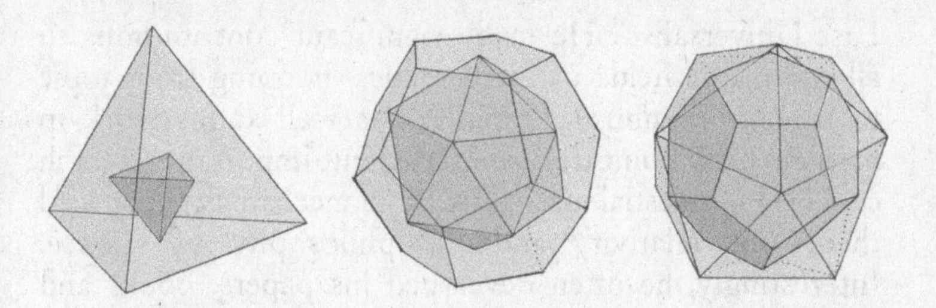

The concept of duals is well worth investigating, as can be appreciated by a simple thought experiment. If you pick up a regular cube, there are certain ways that you can rotate it around so that when you place it back, it looks just like it did before you moved it. You can also reflect it in a mirror, and get something that looks like the original. In other words, cubes possess certain forms of symmetry. The same can be said of the octahedron, and when you move the cube to produce an apparently identical form, the octahedron inside it experiences an equivalent transformation. Because of their connection as duals, it follows that every symmetry of the cube is also a symmetry of the octahedron, and vice versa. Similarly, the dodecahedron has the same symmetry as the icosahedron.

This shows that the five Platonic solids fall into three categories of completely regular three-dimensional symmetry. We shall return to the concept of symmetry in the next chapter, where I examine the advent of non-Euclidean geometries, and the resulting shift in our conception of maths. First, let's have a quick look at the basics of modern topology, and see how Euler's ideas were developed by Henri Poincaré (1854–1912), and the less well-known Simon Lhuilier (1750–1840).

Poincaré and the Birth of Topology

The great French mathematician Henri Poincaré was a worthy successor to Euler. He was a successful writer of popular science books, and is sometimes referred to as 'the

Last Universalist'. He made significant contributions to all the major fields of mathematics, hopping from topic to topic with unusual rapidity. As well as his work in mathematics, Poincaré also carried out important research concerning celestial mechanics, fluid mechanics, the special theory of relativity and the philosophy of science. Interestingly, he often developed his papers, books and lectures by laying out the simplest, most basic concepts he could think of, preferring to start once again from the beginning, rather than jumping into a complex, cutting-edge discussion with experts from each field.

In 1895 Poincaré published a book called *Analysis Situ*, which was an early systematic treatment of topology, particularly the fundamental concept of a 'continuous function'. I won't describe the formal definition of a continuous function, but, as the term suggests, it is the nature of continuous things to have no gaps or sudden jumps. A line is said to be continuous if you can draw it without taking your pencil off the paper. Similarly, we can imagine a continuous function as a rule for gradually distorting a shape, without any tears or jumps. We can, for example, continuously morph a cube into a sphere, or continuously morph a map of Königsberg into our simplified network. When we are looking at the topology of solid shapes, it is rather like the shapes are made from an infinitely stretchy, shrinkable piece of rubber. For this reason, topology is sometimes known as 'rubber-sheet geometry'.

We have already seen that for the Platonic solids, the Euler number is $P-L+R=2$. This result was known to both Euler and Descartes, and Euler's proof of 1752 showed that this result is very general indeed: all manner of faceted shapes have an Euler number 2. However, not every shape has an Euler number 2. It was Poincaré who identified the two topological properties that are logically equivalent to having an Euler number 2:

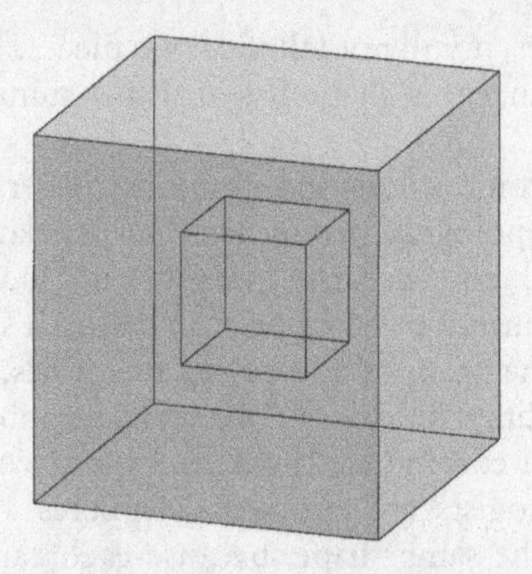

Property 1: All the surfaces of the shape must be connected. Not every shape has property 1. For example, consider a cube with a smaller cube removed from its interior. This shape has an Euler number 4, as the surface of this shape is composed of two disconnected parts, each of which makes a contribution of 2 to the total Euler number.

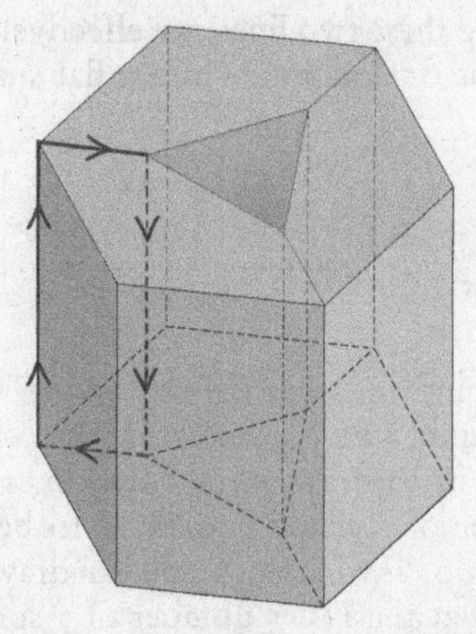

Property 2: If we cut the shape, it should fall into two pieces. The above drawing illustrates a shape that does not have property 2. If we cut this shape along the marked

loop of edges, it will not fall into two pieces. This is related to the fact that this shape has an Euler number of 0.

All the Platonic solids and countless other shapes share these two topological properties. This is related to the fact that we can continuously morph one such shape into another. Continuous functions can distort a shape in many ways, but they cannot add or remove holes. This follows because making a hole requires moving neighbouring points apart, which contradicts the definition of continuous. As far as topologists are concerned, spheres and cubes are effectively the same shape, because each can be continuously distorted into the other. For this reason, we say that spheres and cubes have the same topological identity. Similarly, a doughnut has different topological properties to a sphere, but a doughnut and a coffee cup have the same topological identity. To understand this idea a little better, consider what happens when we draw a network on a doughnut shape.

By drawing these two lines, we effectively cut our shape (called a torus or doughnut) into a flat surface:

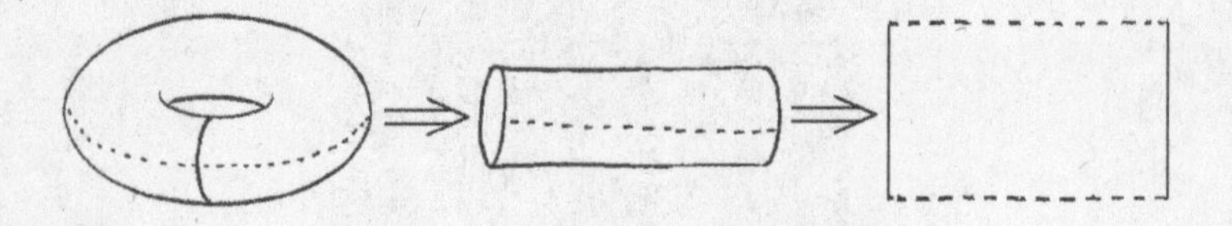

From this point on, steps (I), (II) and (III) cannot affect the Euler number of our network. In other words, the hole enables us to draw exactly two lines 'for free'. The Euler number of a shape is defined to be the minimum Euler number of any network you can draw on that shape. We can see that the Euler number of a sphere is 2, while the Euler number of a torus is 0.

The first general equation concerning topological identities was called 'Euler's Polyhedral Formula'. Despite the name, this formula was first written down by the Huguenot Simon Lhuilier, in 1813. When he was a young man, Lhuilier declined a large fortune because he wanted to study maths, and didn't want the career in the church that was a precondition for receiving his relative's cash. Lhuilier toyed with Euler's work for most of his life, and was taught mathematics by a student of Euler's. He was particularly interested in the argument concerning the bridges of Königsberg, and one of the highlights of Lhuilier's career was spotting a relationship between the Euler number of any shape and the 'genus' of any shape.

Intuitively speaking, the genus of a shape is just the number of holes it contains. So, for example, a sphere has genus 0, while doughnuts and coffee cups both have genus 1. More specifically, we say that a shape has genus 0 if and only if any loop on the surface of the shape defines two separate regions: 'inside' the loop and 'outside' the loop. Notice that a doughnut does not have genus 0, because it is possible to draw a loop on the surface of a doughnut and still leave a single, connected region.

Also notice that we can construct a shape of genus 2 by gluing together two shapes of genus 1, and this process can be extended in the obvious fashion:

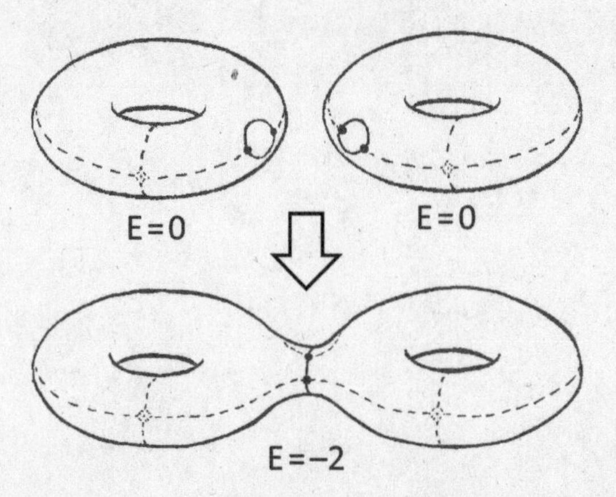

Examine the pieces we are gluing together (the circles made from two points and two lines that can be found on each of the doughnuts). The four lines, four points and two regions of the first drawing make a contribution of 2 to the total Euler number. The corresponding piece of the second drawing consists of two lines, two points and no regions. This is true because the glued part is inside our new shape, so it does not count as a region. Two lines, two points and no regions makes a 0 contribution to the Euler number, which is two less than we had before. In other words, given two shapes with Euler numbers E(1) and E(2), we can glue them together to produce a shape with Euler number $E(1)+E(2)-2$. It follows that gluing a doughnut onto any shape reduces the Euler number by two, and in general the Euler number E is related to the genus g by Euler's Polyhedral Formula, $E=2-2g$.

Chapter 7: EUCLID'S FIFTH AND THE REINVENTION OF GEOMETRY

> 'Insofar as the propositions of mathematics give an account of reality, they are not certain; and insofar as they are certain they do not describe reality. But it is, on the other hand, certain that mathematics in general and geometry in particular owe their existence to our need to learn something about the properties of real objects.'
>
> Albert Einstein, 1879–1955

Measurement and Direction

The first age of geometry can be characterized by the Ancient Egyptians, and their ritual of replacing boundary lines on the flood plains of the Nile. The second age of geometry can be characterized by Euclid, proving geometric truths as a consequence of carefully stated axioms. The third age of geometry did not dawn until the nineteenth century, and in this chapter we will see how mathematicians made this radical shift.

The geometry of 1800 was remarkably similar to the geometry of Euclid, and up until the mid-nineteenth century, geometric statements were understood as being true descriptions of ideal objects in actual, physical space.

Progress had been made, but Euclid's *Elements* was still the major textbook for students of the subject, and in comparison to the other branches of mathematics, geometric arguments were much as they had always been. Indeed, arguments that developed in the newer fields of analysis and calculus were often translated into geometric problems, precisely because geometric arguments were seen as trusty, time honoured and irreproachably sound.

The first sign of a radical break came from long-standing arguments about the proper role of algebraic symbols in geometric proof. However, our conception of geometry did not undergo a fundamental change until people began to explore the mathematics of 'projective geometry', 'non-Euclidean geometry' and 'curved spaces' (or manifolds). To understand the nature of these radical innovations, and the corresponding shift in our conception of geometry, we must first recall the following five axioms, which Euclid used as the basis for all his geometric deductions:

1. There is precisely one shortest path (or straight line segment) connecting any two points.
2. Any straight line segment can be extended indefinitely, forming a straight line.
3. Every straight line segment can be used to define a circle. One end of the segment is the centre of the circle, and its length forms the radius.
4. All right angles are essentially identical, in that any right angle can be rotated and moved to coincide with any other.
5. Given any straight line and a point that is not on that line, there is precisely one straight line that passes through the point, and does not intersect the line.

These statements (plus some other, basic logic) define the concepts of Euclidean geometry. They are related to one another, and together they form a conceptual scheme. Formally speaking, all five axioms share the same logical status. That is to say, within Euclidean geometry, all five axioms are definitively true. Nevertheless, historically speaking, the first four statements are closer to the bedrock of humanity's geometric understanding. If you spend your time using a set square and a tape measure, the first four statements are obviously true. We can demonstrate that the first four statements describe the world of set squares and tape measures, but it isn't clear how you can practically show the reality of Euclid's fifth. Clearly, we cannot physically examine two infinite lines, and confirm that they don't intersect!

In the Introduction, I presented a proof of Pythagoras' Theorem. For those who have eyes to see, this kind of informal proof is utterly convincing. Indeed, an understanding of the truth of Pythagoras' Theorem predates Euclid's system of definitions by many centuries. Given that this is so, it seems fair to say that the Ancient Greeks articulated Euclid's fifth, and drilled it into future generations because of two, intimately related reasons: they saw that it is 'true', and it enabled them to *prove* Pythagoras' Theorem, and other, basic results.

The fifth axiom listed earlier is more accurately referred to as the 'Playfair axiom', after the Scottish mathematician John Playfair (1748–1819). Euclid's original fifth axiom was quite enough to make one's eyes glaze over, as Playfair stated that: 'If a line A and a line B intersect a third line C in such a way that the angle from A to C plus the angle from C to B is less than two right angles, then the lines A and B must intersect. Furthermore, this point of intersection will be on that side of the line C where the inner angles add up to less than two right angle[s], and not on the side where they add up to more than two right angles.'

Euclid's fifth axiom was sufficiently messy to repel generations of mathematicians. Some, like Playfair, found more elegant formulations. For example, the great English mathematician John Wallis (1616–1703) showed that Euclid's fifth is logically equivalent to saying that any triangle can be blown up or shrunk (e.g. by halving the length of each side), without changing any of the angles. Other mathematicians tried to do away with Euclid's fifth altogether by proving this statement using axioms (I) – (IV) As we shall see, that task is provably impossible. Now, when I said that Playfair found a more elegant formulation, I meant that Euclid's original fifth axiom is logically equivalent to the Playfair axiom, so we can happily make a switch. That is to say, we can use Euclid's five axioms to prove the Playfair axiom, and, conversely, if we assume axioms (I) – (IV) together with the Playfair axiom, we can prove Euclid's fifth. The following statement is another logically equivalent alternative to Euclid's fifth: Straight lines parallel to the same straight line are parallel to one another. As any of these choices of axiom enables us to prove exactly the same theorems, our list of defining axioms is somewhat arbitrary, but aesthetic considerations have meant that Playfair's statement is considered the standard form.

As well as providing an axiomatic basis for their geometric deductions, the Ancient Greeks tried to explain the meaning of words such as 'point' and 'line'. For example, Euclid says things like: 'A point is that which has no part'; 'A path is that which is traced out by a moving point'; and 'A line is a breadth-less length'. We might find such comments intuitively vivid, as they paint a picture of the things that we are talking about. Nevertheless, we really don't need to worry about defining our most basic terms since *their mathematical meaning can be fixed by our axioms*, or by any other set of logically equivalent axioms. After all, it is precisely our axioms that tell us what we are entitled to deduce about the things that our proofs are about. To put it another way, there is no point in defining a basic word if our proofs do not actually make use of the definition. A definition other than that given by the axioms might help students see the kind of thing that the teacher wants to reason about, and if we really wanted to we could use more words to define the words that we use in our definitions, but as Blaise Pascal advised in *The Art of Persuasion*, 'Do not try to define anything so obvious in itself that there are no clearer terms in which to explain it.'

Among other things, Euclid's five axioms define our commitments when we say 'these lines are parallel'. We could also say that Euclid's fifth relates to the concept of direction, since we commonly imagine that parallel lines 'point in the same direction'. After two thousand years of following Euclid's scheme, people finally realized that Euclid's implicit definition of direction is not the only mathematically valid one, and our ordinary notions of straight lines and so forth are not the only things that are consistent with axioms (I) – (IV). As we shall soon see, we can work equally well by accepting axioms (I) – (IV) together with a completely different alternative to Euclid's fifth. Furthermore, the various geometric deductions that people had made without referring to Euclid's fifth now make sense (and hold true) in a context that is far broader than was previously appreciated.

Rather poetically, this broadening of the horizons of geometry first occurred in a prisoner-of-war camp. When Napoleon was forced out of Russia, the Frenchman Jean-Victor Poncelet (1788–1867) was captured from among the dead, and marched for five months to a camp on the Volga. He spent two years in Russia, and by the time he was allowed back to France, in September 1814, Poncelet had completed his *Treatise on the Projective Properties of Figures*. I am not going to discuss projective geometry in this book, but I will describe the kinds of non-Euclidean geometries that were independently developed by Carl Friedrich Gauss, Nikolai Ivanovich Lobachevsky, János Bolyai and Ferdinand Karl Schweikart.

Carl Friedrich Gauss (1777–1855) never published his early thoughts on the subject, but as early as 1824 he wrote a letter to a friend that included the following sentence: 'The assumption that the sum of the three angles in a triangle is less than 180° leads to a curious geometry, quite different from ours but thoroughly consistent ...' The first publication of the idea that Euclid's fifth was somehow optional came in 1829, when Nikolai Ivanovich Lobachevsky (1792–1856) presented a paper that a provincial journal agreed to publish, after the St Petersburg Academy of Science rejected the paper as 'outrageous'. Although a number of mathematicians toyed with the idea, the notion of non-Euclidean geometries didn't really become respectable until the 1850s.

Indeed, we might date the change rather precisely, to 10 June 1854, which is when Gauss asked the 27-year-old Bernhard Riemann (1826–1866) to deliver a lecture entitled 'On the Hypotheses that Lie at the Foundations of Geometry'. As we shall see later, the full significance of this lecture wasn't widely appreciated until the work of Einstein, but Gauss certainly enjoyed it. The rock of Riemann's argument was the observation that the basis of geometry is not an intuitive or conceptually necessary knowledge about 'universal space', but simply the capacity to speak properly about *measurements*. Where we measure and formally speak about the lengths and angles of lines on a flat surface, Euclidean geometry is just the tool. Where we measure lengths and angles on a curved surface, we are still doing geometry, but it is no longer 'Euclidean'.

Given that the word geometry derives from the Greek for 'earth-measuring', the foundational significance of measuring lengths and angles may sound ludicrously obvious. However, since the Greeks, mathematicians have had the aristocratic pleasure of knowing all the facts about angles in a triangle and so forth, without having to use anything so crude as an actual, physical ruler. To reassert the appropriate priority of metric theory required a delicate skill: pruning back to the essentials, so that further branches might grow. The elderly Gauss was absolutely delighted by Riemann's approach, and spoke with uncharacteristic enthusiasm about the depth of Riemann's thoughts. Gauss coined the term 'non-Euclidean geometry', which Riemann was too modest to announce, and from that time non-Euclidean geometries were no longer considered as logically curious monstrosities. Instead, the truths of non-Euclidean geometry were accepted along with Euclid's, and the third age of geometry was born.

Non-Euclidean Geometry

By definition, a line between two points is 'straight' if and only if it is as short as any other line that might connect the two points. This fact is fundamental to the meaning of the term 'straight', and it is true in all geometries, not just Euclidean geometry. Euler realized that you can find the shortest path between two points on any convex surface by making a model, poking holes at the two points in question, threading through a length of string and pulling it taut. Because pulling the string makes it as short as possible, it follows a shortest path or 'geodesic'.

The shortest paths between points on a sphere are always 'great circles', which is the name given to circles with the same radius as the sphere in question. If you head straight towards a particular point (e.g. the North Pole), and just keep on going, your path around the earth will draw out a great circle. If two people both head north, their paths cross at the North Pole. Hence the direction 'North' is not what Euclid had in mind while he was writing the fifth axiom, because he specified that parallel lines never cross. Nevertheless, we can produce a perfectly consistent non-Euclidean geometry by reinterpreting our basic terms like 'point' and 'straight line'. We are free to do this because the mathematical meaning of these terms is specified by the axioms alone, and not any other intuitions we may have about the perceived subject matter.

In particular, we can picture a kind of geometry known as 'circular elliptic geometry'. We do this by interpreting the term 'point' to mean 'a pair of antipodal points on a sphere' (that is, points directly opposite one another), 'straight line' as meaning 'great circle', and so on. Axiom (I) is perfectly compatible with this new conception of direction, because there can only be one great circle through any two pairs of antipodal points (and this really is the shortest path):

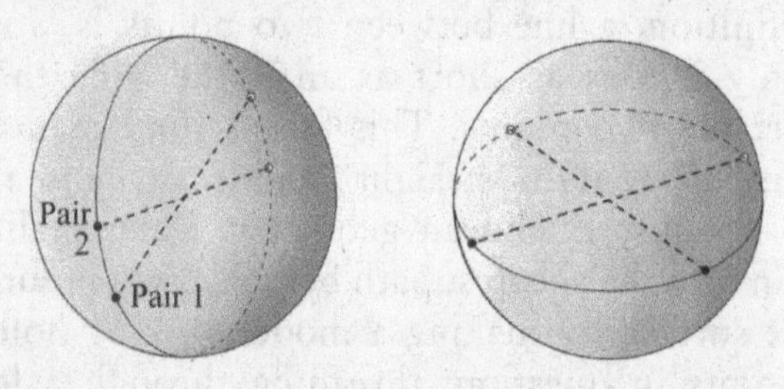

Similarly, axiom (II) is satisfied because any shortest path connecting a pair of points on a sphere can be extended to form a great circle. Axioms (III) and (IV) are also satisfied in this model: we simply interpret the terms 'right angle' and 'circle' by reading them as 'right angle on a sphere' and 'circle on a sphere'. The only somewhat technical problem is that because our sphere is finite, there is an upper limit on the size of line or circle we can draw. This limitation of our model can be overcome, but there is another fundamental difference between this new kind of geometry on a sphere and traditional, Euclidean geometry.

The fundamental difference is that if we draw a great circle and a point not on that circle, we cannot draw any great circle through that point without crossing our original circle. In other words, under this novel interpretation of the terms 'point', 'line' and 'circle', axiom (V) is utterly false, as in this geometry there are no such things as parallel lines. Indeed, if we want to do elliptic geometry instead of Euclidean geometry, we have to adopt the following axiom:

> Given any straight line and a point not on that line, there are no straight lines which pass through the point, but do not intersect the line.

The fundamental point is that we are free to interpret the basic terms 'point', 'line' and 'circle' however we like, so long as our interpretation satisfies our axioms. After all, we are supposed to be relying on nothing but the axioms when making our deductions. Furthermore, we can use the axioms of Euclidean geometry to prove that lines on the surface of a sphere satisfy the five axioms of elliptic geometry. This tells us that elliptic geometry is at least as consistent as Euclidean geometry, as there is a Euclidean object that satisfies the axioms in question.

This argument shows that when people tried to use axioms (I) to (IV) to prove Euclid's fifth, they were attempting the impossible. Euclid's fifth is consistent with axioms (I)-(IV), but so are statements that contradict Euclid's fifth. It follows that if axioms (I)-(IV) are consistent (which they surely are), Euclid's fifth cannot possibly be a logical consequence of axioms (I)-(IV). That is to say, Euclid's fifth tells us something that we simply cannot deduce from the other axioms.

The Curvature of Space

Familiar terms such as 'triangle' and 'square' are perfectly well defined in elliptic geometry:

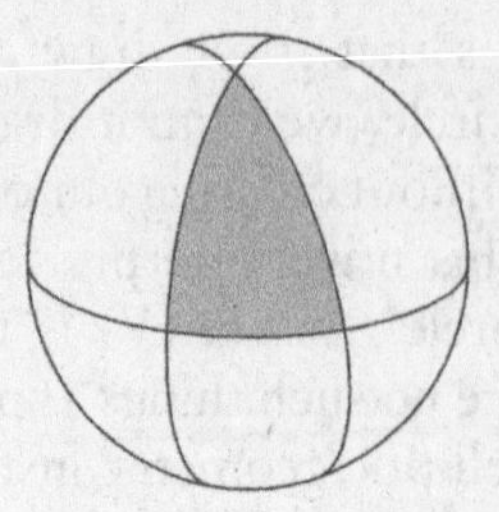

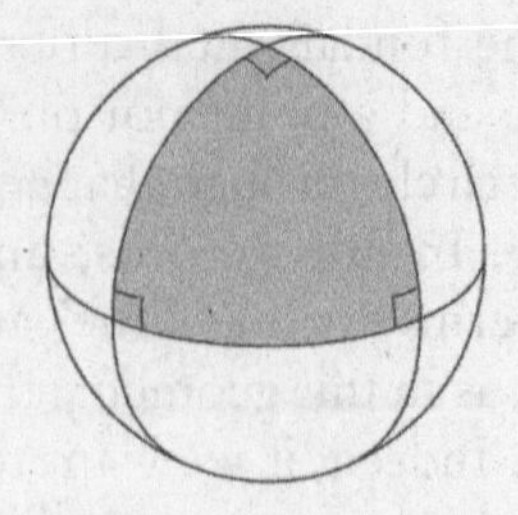

If we add up the angles in the corners of these triangles, we find that the total is more than 180°. The sum of the internal angles is greater than 180° because the boundary lines bulge out, increasing the angles at each of the corners. Triangles that cover a small proportion of a sphere have a sum of internal angles that is very close to 180°, as small parts of a sphere are very close to being flat. For example, the surface of a tub of water is not quite flat as it follows the curvature of the Earth, but this curvature is undetectably small unless the tub is enormous. On the other hand, if we draw a triangle that covers a significant proportion of a sphere, the sum of the internal angles will be significantly larger than 180°. Indeed, our second example of a triangle on a sphere contains three right angles, which gives us a sum of 270°. Hence it is possible to detect curvature by measuring the angles in a triangle, and when we change the size of a triangle on a sphere, the angles in that triangle change.

It was Gauss who first developed a rigorous definition of curvature. The definition of the curvature of a line is really very simple. It rests on the fact that for any three points on a plane, there is a unique circle that passes through all three points (unless all three points lie on a straight line).

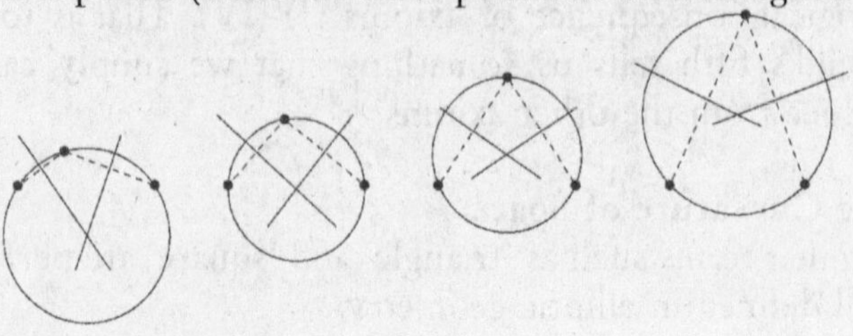

The curvature of a straight line is said to be zero, while the curvature of a circular arc of radius r is said to be $1/r$. Given a smooth, continuous line, we define the curvature of each point p by finding a 'limit case': an idea that relates to the infinitesimal calculus. Essentially, Gauss imagined a line through a point p, together with a point on the line before p and another point on the line after p. There is precisely one circle that passes through any three points, and we can find the value of $1/r$, where r is the radius of the circle that passes through all three points. If we pick a closer pair of points before and after p, we get another value of $1/r$. In the case of smooth curves with no corners, the value of $1/r$ gets closer and closer to some fixed value as the points before and after p move closer and closer to p. By definition, we say that the curvature at p is the limit case $1/r$.

More importantly, Gauss also developed a definition for the curvature of a surface, which relies on nothing more than measurements of angle and distance within the surface itself (that is to say, Gaussian curvature does not depend on the space in which our surface is embedded). The basic idea is that if we move along a surface, our line of motion will have a curvature. If we start at a given point and move in different directions, we will find lines of different curvature, and in particular we will find one direction of travel with the smallest curvature k_1, and another direction of travel with the largest curvature k_2. We also say that these curvatures are positive if we are on the top of a hill, and negative if we are at the lowest point of a bowl. Now, by definition, the Gaussian curvature at the given point on a surface is equal to k_1k_2. Note that the top of a hill and the bottom of a bowl both have positive curvatures, as k_1k_2 is positive if k_1 and k_2 are both negative. Also note that if either k_1 or k_2 is zero, the Gaussian curvature is zero. Finally, note that some surfaces have a negative curvature, as in the case of a 'saddle point', where you move downhill if you go forwards or backwards, but you move uphill if you go left or right.

If a surface contains a (Euclidean) straight line, then the points on that line must have a Gaussian curvature of zero. For example, a cylinder and a cone both have Gaussian curvature zero, as at each point the path with the least curvature is a straight line, so we have $k_1 = 0$, which means that $k_1k_2 = 0$. In other words, we can identify shapes with zero Gaussian curvature by placing them on a flat surface. If there is a line of contact between the shape and the flat surface, the surface must have zero Gaussian curvature. If there is a point of contact (as we find in the case of a sphere), the surface does not have zero Gaussian curvature. Another way to identify surfaces with zero Gaussian curvature is to draw a triangle, taking the shortest path along the surface between three given points. If the surface has zero Gaussian curvature the internal angles of this shape will add up to 180°. If the surface has positive Gaussian curvature the internal angles will add up to more than 180°, and if it has negative curvature the angles will add up to less than 180°.

Surfaces with zero Gaussian curvature can be rolled out flat, preserving all the geometric relationships between the points on that surface, while surfaces with non-zero Gaussian curvature cannot be rolled flat. For example, a cylindrical roller with an engraved geometric pattern works just fine, and can be used to transfer a pattern from the roller onto a flat piece of paper. In contrast, the curvature of a sphere means that it cannot be used as a roller, and any flat map of the globe necessarily distorts at least one geometric relationship. That is to say, it is provably impossible to map the surface of a sphere onto a flat surface without distorting either the lengths or the angles between points on the surface of the sphere. At best you can produce a map that distorts the lengths but not the angles, or the angles but not the areas.

Gauss called his insight into curvature the *Theorema Egregium* (Latin for 'remarkable theorem'), and it explains the effectiveness of a common pizza-eating strategy. Because it is basically flat, a slice of pizza can be seen as a surface with Gaussian curvature zero. You can bend and fold a slice of pizza, but it is difficult to stretch, and that means it is difficult to change its Gaussian curvature. Similarly, you can bend or fold a piece of paper, but because it is difficult to stretch a piece of paper, you can't really wrap a spherical present with flat wrapping paper. Now, imagine wrapping a slice of pizza around an invisible cylinder. In one direction the pizza is curved like a circle, but because the pizza has zero curvature, in the perpendicular direction it has to lie flat. The slice of pizza cannot bend in both directions without having a non-zero Gaussian curvature, and it cannot change its curvature without stretching. In other words, folding a pizza creates rigidity in the direction perpendicular to the fold, which is useful because it stops the pizza flopping, and keeps the toppings nicely in place. The same kind of reasoning also explains why a corrugated sheet of metal is nowhere near as floppy as a flat sheet of the same thickness.

While Gauss was the dominant figure in mathematics, the remarkable French mathematician Marie-Sophie Germain (1776–1831) made a number of significant contributions, and in 1816 she became the first woman to win the *Grand Prix* of the Paris Academy of Sciences. One of her most brilliant ideas was to realize that Gaussian curvature isn't the only useful definition of curvature: the quantity $\frac{1}{2}(k_1 + k_2)$ is also very informative. She called this quantity the mean curvature of a surface (at the given point), and she noted that the mean curvature of a surface has a fascinating connection with its surface area. More specifically, if a surface has zero mean curvature at every point, you cannot distort that surface without increasing its surface area. This means that materials that tend to minimize their surface area will naturally form minimal, zero mean curvature surfaces. For example, it is a physical fact that if you dip any shaped piece of wire into some soapy water, the surface tension pulls the surface area to a minimum. A soap bubble contains a volume of slightly compressed air, and it minimizes its surface area by making a sphere (a shape that has the same curvature at every point on its surface). The air inside a soap bubble is at a slightly higher pressure than the air outside the bubble, but when a soap film is subject to a single pressure, every point on its surface must have a mean curvature of zero. This means that if we dip some wire into soapy water, the soap film that spans the space between the wires must have zero curvature, however we bend the wire.

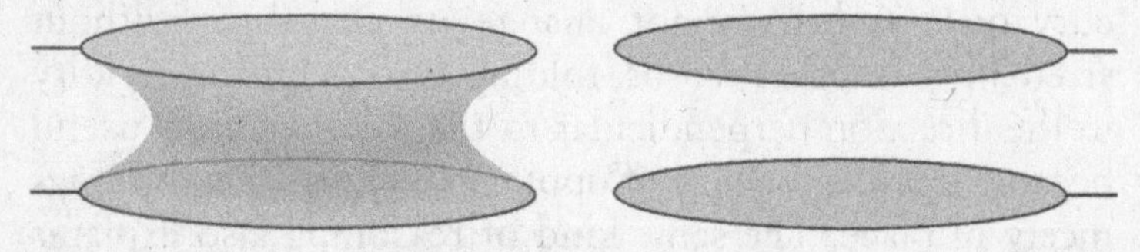

Also note that a soap film can span between the same wire shapes in more than one way, though each point on the surface will always have a mean curvature of zero.

The Unity and Multiplicity of Geometry

As well as the positive curvature of a sphere's surface and the associated elliptic geometry, there is also something called 'hyperbolic geometry'. In other words, geometricians can also work in a space with negative Gaussian curvature. In hyperbolic geometry the angles in a triangle add up to less than 180°, but the most fundamental fact is that we work with the following axiom:

> Given any straight line and a point not on that line, there are at least two straight lines which pass through the point, but do not intersect the line.

Hyperbolic geometry is somewhat less intuitive than elliptic geometry, but surfaces with negative curvature do occur in nature. For example, coral reefs often have negative Gaussian curvature, but given that the following illustrations need to appear on a flat page, it is easiest to use the model of hyperbolic space devised by Henri Poincaré. On Poincaré's circular map, objects in the space appear to shrink as they move closer to the edge, but according to the internal measure of hyperbolic space (which is the only thing that matters), there is an infinite distance from the centre of the map to the edge. In other words the circular boundary is not a location within the space itself, as for the inhabitants of the hyperbolic world it is an infinite distance away.

A full account of hyperbolic geometry is beyond the scope of this book, but Poincaré's map conveys the basic idea. On this map, a 'straight line' is either a diameter across the circular map, or a circular arc that meets the map's boundary at right angles. Two lines are said to be 'parallel' if they intersect on the boundary. Finally, circles in hyperbolic space look like ordinary circles, but their centres are not where one would naively expect. Poincaré's map accurately represents the angles of hyperbolic space, so we can see that the angles of a triangle add up to less than 180°, while the angles of a quadrilateral add up to less than 360°:

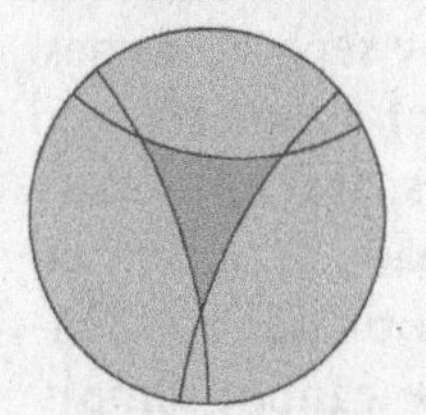
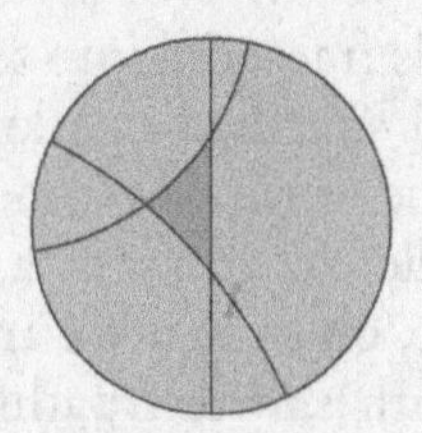
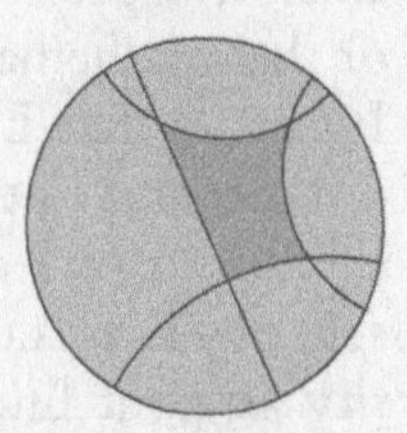

The significance of there being more than one geometry is frequently misunderstood, but the development of non-Euclidean geometry triggered major changes in the public perception of maths. Statements about the factual world of countable objects and measured spaces lie at the heart of mathematics. Indeed, we might say that the language of facts gives us the language of mathematics. It is an ancient knowledge to describe a square field as a mathematical square, and the truths of geometry are certainly facts of our world.

However, by the end of the nineteenth century, non-Euclidean geometry had helped to create a popular image of 'pure mathematics': a logical, deductive discipline disassociated from worldly facts. The ancient technique of *reductio ad absurdum* involves the description of things that aren't true, where you state a proposition and conclude that you must reject it as false. What was new was the notion of accepting and exploring axioms 'whether or not they are true'. This was a radical and disruptive idea, and it changed the public perception of the nature of mathematics. In particular, it gave rise to the idea that only some mathematics was concerned with the real world, as up until the late nineteenth century, absolutely no one divided mathematics into 'pure' and 'applied'. In the final chapters of this book I shall reconsider the controversial relationship between mathematics and the physical world. First I want to discuss 'absolute geometry', and the significance of having more than one form of geometry.

The first point to make is that the existence of non-Euclidean geometries does not mean that Euclid was wrong, or that mathematics is fractured into totally separate types. It is true that Euclid specified one kind of geometry and not others, but these various geometries are intimately related to one another as parts of a broader conceptual scheme. For example, given that we are motivated to identify straight lines with shortest paths, we cannot simply draw a squiggle and claim that it defines a 'direction', or some new kind of geometry. We have to have a measure by which our 'straight lines' are 'shorter' than every other way of moving between two points.

Because Euclid and his followers were careful to avoid Euclid's fifth whenever possible, many theorems apply to any reasonable definition of direction. For example, if one rotates a match through the internal angles of a three-sided figure, its head switches side as it rotates through half a loop. If you rotate through the internal angles of a four-sided figure, the head of the match does not switch sides as it rotates through a complete loop.

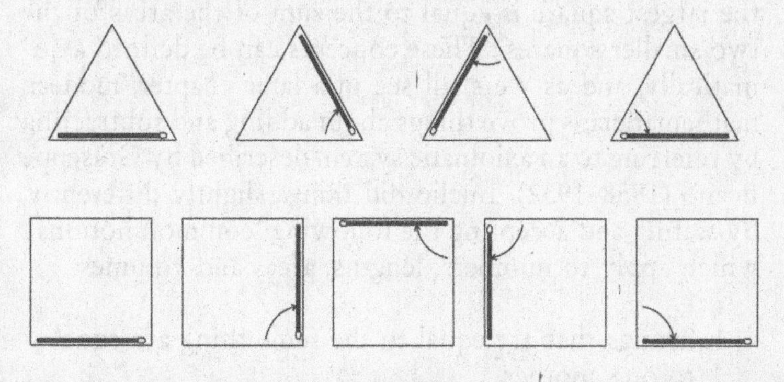

This observation is said to be part of absolute geometry, because it can be proved on the basis of axioms (I) – (IV) alone. This means that it holds true regardless of curvature, in Euclidean, elliptic and hyperbolic geometry. On the other hand, it is a distinctive characteristic of Euclidean geometry that if we measure the angles at the corners of these shapes, the total comes to 180° in the case of triangles, 360° in the case of quadrilaterals, and more generally the internal angles of an N-sided polygon will total $180^\circ \times (N - 2)$.

Each axiom operates in the context of the other axioms, all of which contribute to the meaning of our words or symbols. The way that axioms work together is absolutely fundamental, and logical terminology plays an essential role in enabling an axiom to be descriptive of a mathematical concept. Mathematical logic is a subject we will return to. At this point I just want to point out that in order to make sense, collections of axioms must cohere as a system. In particular, if we are to work from the axioms to their logical consequences, the axioms must contain words such as 'and', 'or', 'not', 'every' and 'some', which can support logical deductions.

Euclid's arguments are definitively geometric when they refer to his five axioms. His arguments also involved the concepts of 'equals', 'plus' and 'minus'. For example, in a proof of Pythagoras' Theorem we might say 'the area of the largest square is equal to the sum of the areas of the two smaller squares'. These concepts can be defined axiomatically, and as we shall see in a later chapter, modern mathematicians prove things about adding and subtracting by referring to an axiomatic system described by Giuseppe Peano (1858–1932). Euclid did things slightly differently, by stating and accepting the following 'common notions', which apply to numbers, lengths, areas and volumes:

1. Things that are equal to the same thing are equal to one another.
2. If equals be added to equals, the wholes are also equal (e.g. if $A = B$ then $A + C = B + C$).
3. If equals be subtracted from equals, the remainders are also equal.
4. Things that coincide with one another are equal to one another. In other words, shapes that lie on top of one another have the same dimensions.
5. The whole is greater than the part. For example, the area of a shape must be at least as large as the area of any part of that shape.

As a final comment on the significance of multiple geometries, let's return to Riemann's lecture 'On the Hypotheses that Lie at the Foundations of Geometry'. The first part of his lecture was essentially concerned with geometry itself. The second part of this now famous lecture posed deep questions about geometry in the physical world, asking about the dimension and geometry of actually measured space. This would have been inconceivable in an earlier age, when it was accepted that Euclid's axioms captured an immutable framework for spatial comprehension. In particular, notice that axioms (I)–(IV) are physically highly plausible. We can connect any two points with a taut string (I). We can unroll some more string and extend the line (II). We can use a length of string to define a circle (III), and so on. Moving a physical shape really doesn't change its area, and axioms (I)–(IV) all capture statements as physically obvious as this. Euclid's fifth is the only axiom that we cannot clearly demonstrate using stretched threads or rays of light. This is because it makes a claim about the whole extent of a line, as a pair of parallel lines must never cross, no matter how far they are extended.

Gauss actually bothered to make a physical check on Euclid's fifth, and as far as he could tell, measurements on earth are in agreement with Euclidean geometry. For example, if we measure the angles in a triangle of light between three mountain tops, we cannot detect a difference between the sum of those angles and 180°. At the time, very few people appreciated the significance of the ideas in the second part of Riemann's lecture, but sixty years later Riemann was dramatically vindicated by the general theory of relativity. Crucially, Einstein realized that we can give the word 'straight' a physical definition by saying that the path of light in a vacuum will always be straight (by definition), but his analysis indicated that we cannot find the shortest path without considering the presence of mass. In other words, Einstein predicted that light was affected by gravity, so, in a sense, light rays 'bend' around massive objects. In 1919, during an eclipse, astronomers confirmed that when light from a star passed nearby the sun, the pattern of stars was bent just as Einstein had predicted.

If we want to measure a building on earth or calculate where a rocket will land on the moon, Euclidean geometry is just the tool we need. If we want to calculate the path of light from a distant galaxy, we turn to Einstein's geometry instead. By analogy, we can detect the curvature of a sphere by measuring the angles in a triangle. We just need to draw a large enough triangle. Similarly, the astronomical observations that made Einstein famous detected a slight warping of angles. It isn't easy to get your head around, but we can describe this situation with perfect mathematical accuracy by identifying gravity with a curvature of space-time. In other words, the earth can 'detect' the presence of the sun (and moves accordingly) because space itself is curved. This idea is often represented by the image of a bowling ball on a trampoline. The surface of the trampoline deforms and curves under the weight of the ball, and if we place a marble on the trampoline, it will accelerate towards the ball. Likewise, the presence of mass induces a curvature in space-time, and the effect of that curvature is to produce the acceleration due to gravity that we are all familiar with.

Symmetry and Groups

The naïve concept of symmetry is very ancient indeed, and in ordinary usage, something is said to be symmetrical if its left half is the mirror image of its right half. For several centuries now mathematicians have been rather more precise, but also more general, in their use of the word 'symmetry'. For mathematicians, a symmetry is understood as being a particular kind of transformation, or rule for moving or changing a mathematical object or shape. More specifically, an object is said to possess a particular symmetry if and only if performing the given transformation (or symmetry) produces an object that is apparently identical to the original. So, for example, an object that is bilaterally symmetric (i.e. an object whose left half is the mirror image of its right half) is indeed mathematically symmetric precisely because reflecting the right half onto the left and the left half onto the right yields a shape that is indistinguishable from the original.

One of the big advantages of the mathematical definition of symmetry is that it is extremely general. There are many kinds of transformation other than reflection! For example, it is mathematically useful to observe that every shape has at least one symmetry, namely the identity transformation. That is to say, if you perform the transformation 'leave every point in the shape exactly where it is', the end result is identical to the original shape. A more interesting kind of symmetry is rotational symmetry, which can be found in the following shape:

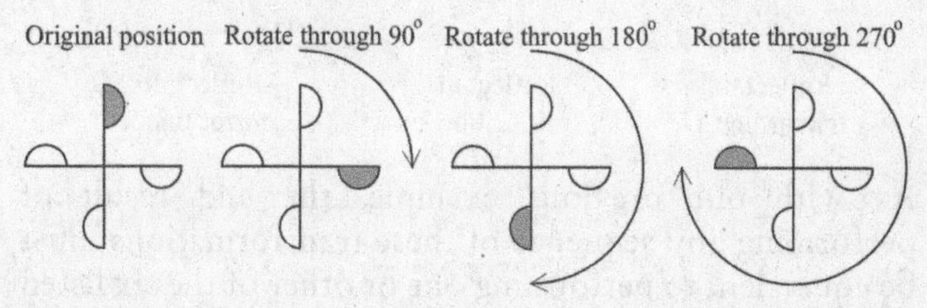

For obvious reasons, this shape is said to possess four-fold rotational symmetry.

Note that if we effect several of these transformations in a row, the net result is equivalent to one or other of the four listed symmetries. If we rotate our shape through 180°, and then rotate it through 90°, the net result is a rotation of 270°. Similarly, if we rotate our shape through 270°, and then rotate it through 180°, the net result is a rotation of 450°, which is equivalent to a rotation of 90°. The basic and fundamental fact is that because each of these transformations produces a shape that is indistinguishable from the original, any combination of these transformations must also produce a shape that is indistinguishable from the original.

To give another example of a symmetric shape, the following is said to possess three-fold rotational symmetry, as well as reflective symmetry:

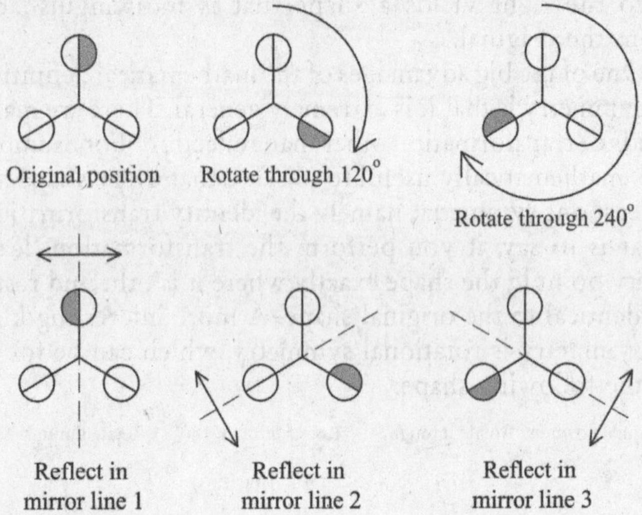

Original position
Rotate through 120°
Rotate through 240°

Reflect in mirror line 1
Reflect in mirror line 2
Reflect in mirror line 3

As with our previous example, the end result of performing any sequence of these transformations must be equivalent to performing one or other of the six listed symmetries. For example, if we reflect our shape in mirror line 1 and then rotate it through 120°, the net result is equivalent to a reflection in mirror line 2. Similarly, if we reflect our shape in mirror line 2 and then reflect in mirror line 1, the net result is equivalent to rotating our shape through 240°.

Reflection and rotation are by no means the only transformations that can be considered as symmetries. Perhaps the simplest example of another kind of symmetry is what mathematicians call 'translation', where we move a given shape in a fixed direction by some fixed amount. For example, if we have an infinitely long string of identical beads we can shift each bead one place to the left and the result is indistinguishable from the original configuration.

Likewise, we could shift each bead one place to the right, two places to the left, seventeen places to the left, and so on, and the result would be indistinguishable from the original configuration.

As we have already seen, modern mathematicians use something known as 'group theory' when they are describing symmetries, and this branch of mathematics has quite justly been described as the supreme art of mathematical abstraction. All groups are comprised of a set of elements together with an 'operation' or rule for combining those elements. To qualify as a group, the elements and operation must satisfy the four, definitive rules listed below. The most familiar example of a group is the integers (our set of elements), together with addition (our rule for combining elements). Another example of a group is the set of symmetries of any object where we combine pairs of symmetries by performing one symmetry transformation followed by the other. Now, the properties that define a group are as follows:

Closure: If we combine any two elements from our set, the result must be another element from our set. For example, if we add any two integers, the result is always another integer. Similarly, recall that the symmetries of a four-fold rotationally symmetric object are 'rotate by 90°', 'rotate by 180°', 'rotate by 270°' and the identity transformation, 'rotate by 0°'. If we combine any two elements by doing one and then the other, the net result will be equivalent to one of our four elements, which means that we have 'closure'.

Associativity: The operation must be associative. In other words, if you combine any three of the elements, it cannot matter whether you combine the first two and then combine the result with the third, or combine the last two and then combine the result with the first. For example, addition is associative, as is shown by the following example:

$(2+5)+1=7+1=8$, and likewise $2+(5+1)=2+6=8$.

The operation of following one transformation by the other is also associative, because, for example:

('rotate by 90°' then 'rotate by 180°') then 'rotate by 90°' = ... 'rotate by 270°' then 'rotate by 90°' =

'rotate by 360°'.

Likewise, 'rotate by 90°' then ('rotate by 180°' then 'rotate by 90°') = 'rotate by 90°' then 'rotate by 270°' =

'rotate by 360°'.

Identity: The group must contain an identity element. In other words there must be one element that, when it is combined with any other element, leaves that element unchanged. In the case of the integers, the identity element is zero, because, for example, $5+0=0+5=5$. In the case of groups of symmetries, the identity transformation is 'leave every point exactly where it is'.

Inverse: Every element in the group must have what is known as an inverse. If you combine an element with its inverse, then by definition the result is the identity element. For example, the inverse of 5 is -5, because $5-5=0$, and zero is the identity element. Similarly, the inverse of 'rotate by 90°' is 'rotate by 270°', because:

'rotate by 90°' then 'rotate by 270°' = 'rotate by 360°' = 'rotate by 0°'.

Note that in some cases an element is the inverse of itself. By definition the identity element has to be its own inverse, and 'rotate by 180°' is another example of an element which is its own inverse.

Group theorists summarize and in a sense completely describe the groups they study by using multiplication tables. These are square tables that contain a row and a column for every member of the group. For example, we look to the space in the row marked *r* and the column marked *s* to find the result of combining *r* with *s*. The point is that the multiplication table tells us everything we need to know about combining the elements of our given group. The axioms of group theory impose constraints on the possible forms that such tables can take. For example, the inverse axiom tells us that by definition, every row and every column of every group's multiplication table must contain the identity element *I*.

The crucial fact is that there are limited numbers of multiplication tables. For example, there is only one abstract group containing three elements (the one whose elements are r, r^2 and $r^3 = I$), although this single kind of group can be illustrated in countless different ways. We might, for example, say that *r* is the transformation 'rotate by 120°'. Another way to consider the same abstract group is to imagine that we have three playing cards in our hand, and then say that *r* represents the operation 'move the first card to the back'. In this case r^2 represents moving the first card to the back and then repeating the operation. As in the case where *r* is 'rotate by 120°', our group of card shuffling operations implies that r^3 is the identity transformation 'leave the cards where they are'. The point of all this is that although our two groups are illustrated in different ways, they have exactly the same multiplication table, and so we say that they are two examples of exactly the same abstract group.

To recap, the different kinds of abstract group can be

distinguished by their multiplication tables, and there is only one multiplication table for groups containing three elements. However, in some cases two completely different groups have exactly the same number of elements. For example, consider the following pair of groups, and note that only in the second group is every element its own inverse:

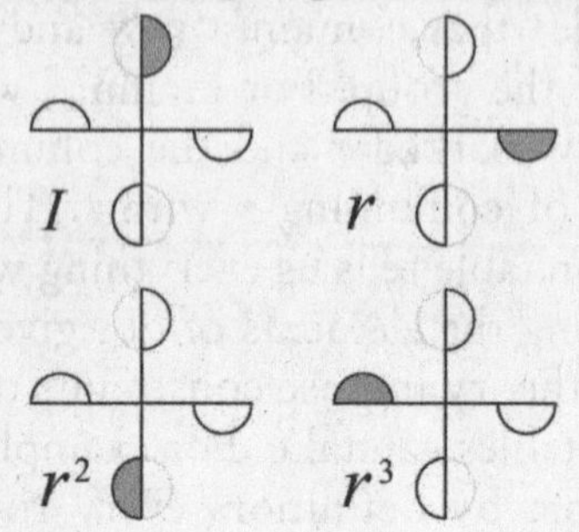

	I	r	r^2	r^3
I	I	r	r^2	r^3
r	r	r^2	r^3	I
r^2	r^2	r^3	I	r
r^3	r^3	I	r	r^2

In this illustration *I* denotes the identity transformation while *r* denotes a rotation of 90°.

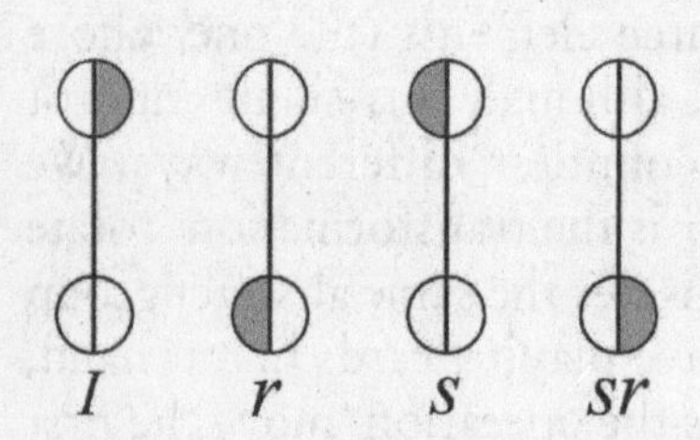

	I	r	s	sr
I	I	r	s	sr
r	r	I	sr	s
s	s	sr	I	r
sr	sr	s	r	I

In this illustration *I* denotes the identity transformation, *r* denotes a rotation of 180° and *s* denotes a reflection in a vertical mirror line.

Both of these groups contain four elements, and so we say that they are both groups of order four. Furthermore, these two groups are the only ones that contain four elements: every other multiplication table with four elements fails to satisfy the axioms of a group. In other words, there are precisely two abstract groups of order four, though each one of these may have countless illustrating instances. Since 1980, mathematicians have had a complete classification of all the different 'simple finite groups'. We know, for example, that if a group contains p elements where p is a prime number, then there is only one possible multiplication table for that group. In fact, there must be some element r such that $r, r^2, r^3, \ldots, r^p = I$ are the only elements of the group. In other words, every group with a prime order p is equivalent to the group generated by r, where r is the transformation 'rotate by $360°/p$'. The situation is more complicated when a group does not have a prime number of elements, as in that case there may or may not be several possibilities for the multiplication table. For example, there are two different groups that contain six elements, five different groups that contain eight elements, but there is only one group that contains fifteen elements.

The Oddities of Left and Right

Mirrors are intriguing objects. Perhaps the strangest thing is that we can measure every facet of an object, from length, surface area, volume, colour or texture, yet none of these will help us to distinguish a left hand from the right. The oddity is that in order to name left and right correctly we cannot look to the object alone. As we shall see, this is because left and right do not carry their differences with them, but standing together we can establish an arbitrary convention, calling this one 'left' and that one 'right'. So why are left and right so similar but different, and why are there two distinct but equivalent options, and not some other number?

There are two fundamental facts that underpin the answer to these questions. The first point to note is that there are two ways to travel along a line: 'forwards' and 'backwards'. The second point to note is that the mathematical operation of moving a shape does not change the object's shape or properties: it only changes the shape's location. As we shall see, these two facts are the key to understanding why mirror images are the same but different.

First, let's consider two-dimensional shapes. If we cut the following shapes out of a sheet of paper and keep them flat on our desk, there is no way that we can move them around so that one is on top of the other. Within the two-dimensional space there are the two distinct shapes: a left-hander and a right. However, if we pick one of them up and turn it over, its front becomes its back, which switches a left-hander to a right (or the other way around).

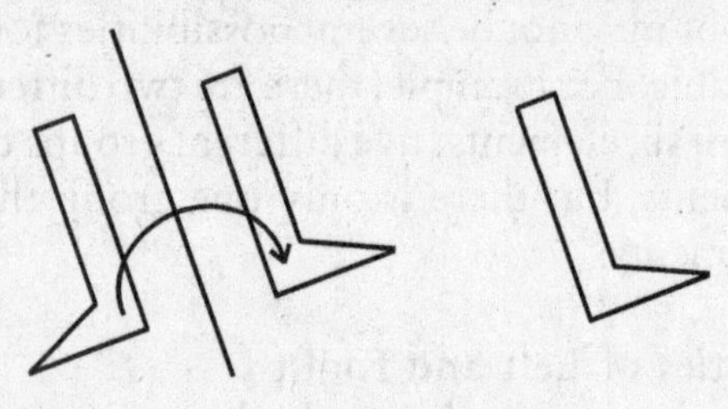

In other words, when we compare the outlines of these shapes within a two-dimensional space, we can see that they are different. One faces left while the other faces right, and no amount of rotation will make the two outlines the same. However, if we compare the outlines within a three-dimensional space we can see that they are identical. When we flip one of them over it looks just like the other, and so by definition the two outlines have the same shape (one is just a rotated version of the other). This observation can help us to understand the rather peculiar way that reflecting an image in a mirror changes the image, but also leaves it the same.

To put it another way, people sometimes wonder why mirrors switch left and right. The answer to this question is that a mirror changes parity (so left-handed gloves look like right-handed gloves), but it is *not* left and right that are switched. Imagine a person standing in front of a mirror, with a glove on their left hand but not on their right. The person's head will be next to their reflection's head, and their gloved hand will be next to their reflection's gloved hand. In other words, the up–down axis and left–right axis remain unchanged. The person differs from their reflection because mirrors switch forwards and backwards, so we stand face-to-face with our mirror image, rather than seeing the back of our reflection's head.

In contrast, imagine dressing two identical twins with gloves on their left hands, and standing them face-to-face. If one twin is facing North the other faces South, and if one has a glove pointing West, the other has a glove pointing East. The only dimension in which both twins are the same is 'up–down', as both twins have their feet on the ground and their head in the air. In other words, a mirror reverses your image in just one dimension ('front–back'), but in the case of face-to-face twins, two dimensions have been switched ('front–back' and 'left–right').

We can change the parity of two-dimensional objects by rotating them in three-dimensional space. Similarly, we can mathematically rotate three-dimensional objects in a four-dimensional space, and when we do that, right-handed objects become left-handed objects. As we cannot directly observe such a transformation, it is tempting to imagine that turning a right-handed glove into a left-handed one requires a moment when the object goes 'pop', as some essence is suddenly switched from type-left to type-right. However, as we will see in the following section, there is no sudden moment when the object is switched. There is nothing truly intrinsic to being left-handed or right-handed, and we find the change by comparing a before-and-after version of the glove, not by looking for a distinctive moment when the essence of our shape is transformed.

The Möbius Strip

Sometimes the mathematical community is ripe for an idea, and more than one person takes the same elementary step. In 1858, Johann Listing and August Ferdinand Möbius independently described the mathematical object now called a 'Möbius strip'. This fascinating shape can be used to elucidate the fundamental equivalence between mirror images, and it is worth the effort of making one yourself. Simply take a strip of paper, twist one end through 180°, and glue it to the other end. This shape has only one edge, as if you run your finger along the side, it twists its way around both top and bottom. If you take a pair of scissors and cut the strip along the middle, a single loop emerges. This new loop has two edges: what was once the middle, and the single edge of the Möbius strip.

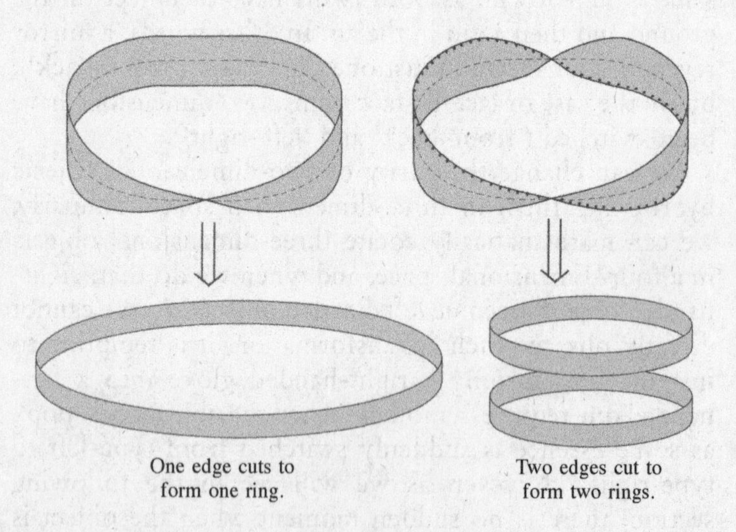

One edge cuts to form one ring.

Two edges cut to form two rings.

We can make a shape called the Möbius strip, but we can also make a space called the Möbius space. We simply take two strips of paper, twist the ends and stick them together. The gap between the strips is a perfectly ordinary two-dimensional portion of three-dimensional space. Mathematicians skilfully ignore the edges of the strip because it is irrelevant to the questions we are interested in asking. We do this by imagining that the 'two' sides are curled up like a cylinder, although it isn't physically possible to join up the edges like this. The important point is to imagine what it would be like for a pair of two-dimensional objects that live in a Möbius world.

On meeting each other, our pair of objects might agree that they are 'mirror images'. Now suppose that one of them sets out for a walk, and comes back to its former acquaintance. This time each one meets their identical image, as a loop through the space effectively switches left and right. It would now appear that they are 'the same shape'.

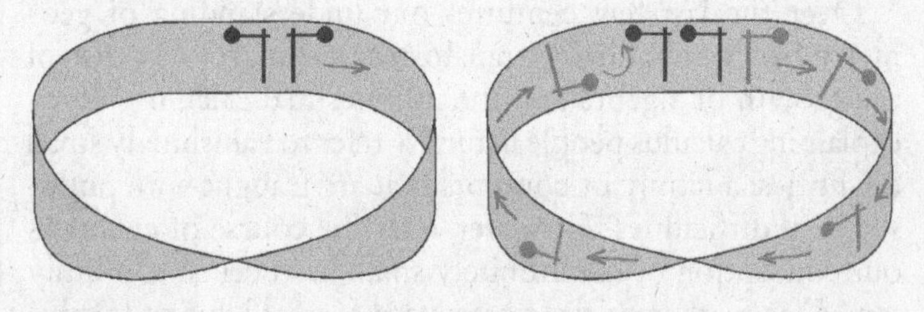

The shape on the right goes for a clockwise walk around the Möbius space, and apparently comes back switched.

The one that has not moved says, 'You have changed – we used to be different but now we are the same.' The other replies, 'I think I would have noticed if a change had taken place. It is you that must have changed.' Now suppose that our two shapes draw identical pictures of themselves, and agree that whoever matches the picture deserves the name 'RIGHT'. Let us follow the fate of the shape on the left as it moves around the Möbius space. It continues to match its picture of 'RIGHT', and finds its way back to its companion.

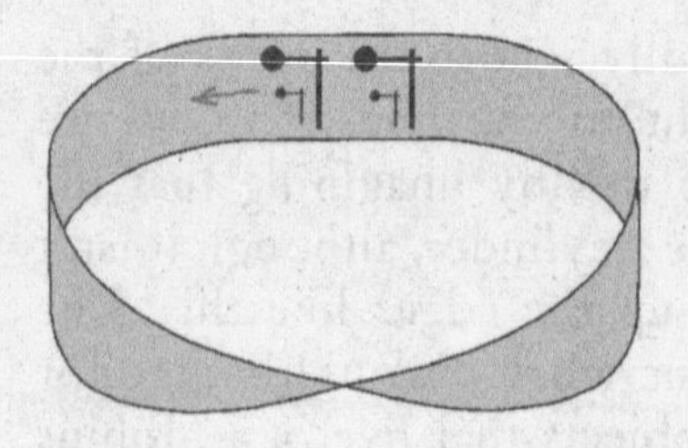

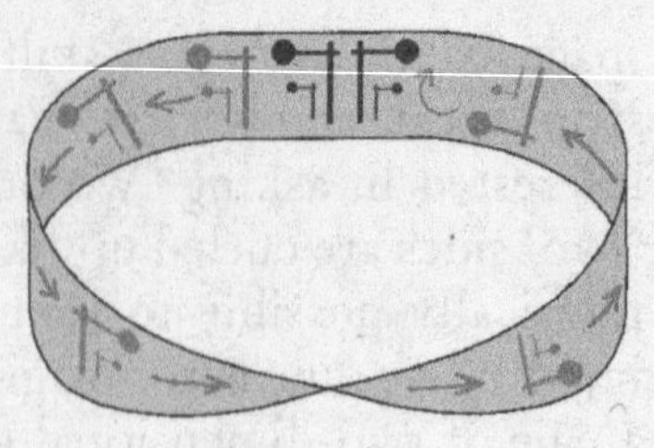

Now they are different again, but both of them are 'RIGHT'. This is because we are mistaken in thinking we have said anything at all when we made our claims about same and different. There is no essence to being of one form or the other – it is in *comparison* that each claims its place (being the 'same' or 'different' to one another). In the Möbius space, left is identical to right, and there simply is no parity to speak of.

Over the last few centuries our understanding of geometry has been radically transformed, primarily because of the growth of algebra and the infinitesimal calculus. When explaining calculus people naturally refer to vanishingly small quantities: a family of concepts that are fraught with philosophical difficulties. However, over the course of centuries our conception of the infinitely small has been successfully tamed, as mathematicians have used formal logic to formulate explicitly the definition of the limit case. In the next chapter we will see how mathematicians have tackled other kinds of mathematical infinity, extending our concept of the infinite, and proving results about infinite sets.

Chapter 8: WORKING WITH THE INFINITE

'What has been said once can always be repeated.'
Zeno of Elea, *c.* 490–430 BC

Blaise Pascal and the Infinite in Maths

Given any number, we can always say 'add one'. What is more, there is nothing to stop us from repeating this instruction. The unbounded or infinite character of the integers is a profound and basic truth: there is no largest number. We can also encounter the infinite by asking, 'How many points on a line?' Given any line we can mathematically chop it in half by identifying the mid-point. This operation leaves us with two ordinary lines, so we can chop and chop again an infinite number of times.

The mathematically infinite was a hot topic of debate for Greek mathematicians and philosophers, as well as later scholars. In particular, Aristotle famously proposed that there are two kinds of infinity: the potentially infinite and the actually infinite. An actual infinity is one that is completed, definite and consists of infinitely many elements. In contrast, a potentially infinite sequence is simply a finite sequence that can be extended indefinitely. So, for example,

Aristotle described a line of finite length as potentially infinite with respect to division, because a line can be divided into two halves, which in turn can be divided into four quarters, and so on *ad infinitum*. To this day many people are prepared to acknowledge only the existence of the potentially infinite, denying the reality of actual infinities.

In the previous chapter we touched on the topic of 'infinitesimals'. That is to say, at various points in history, mathematicians have studied curves by considering an interval that becomes smaller and smaller. This process suggests the idea of an infinitely small interval, but mathematicians do not need a special number that is infinitely small: we simply use the formal notion of a limit case. To put it another way, mathematicians skilfully avoided reference to the infinitely small. As we have seen, the real number 0.999… is equal to 1, and there is no infinitely small gap between them.

In this chapter we will see some other ways that mathematicians have worked in the shadow of the infinite. In particular, we will examine a form of proof first described by Blaise Pascal, which invokes Zeno of Elea's principle of saying something and repeating. We will also see how Georg Cantor established transfinite mathematics, taming the actually infinite as a part of set theory. First, I want to write about my favourite thinker: Blaise Pascal (1623–1662).

Pascal thought very deeply about uncertainty, authority and rules, and his insights permanently changed the way we think about science and religion. No medieval person would question whether belief in Christianity was 'rational' or not, as it was assumed that evidence for the existence of God was all around us. Pascal was a deeply religious man, but he was clear that we are led to Christ by the heart, not by reason, and we should admit that accepting the teachings of the church involves a leap of faith.

Furthermore, Pascal was very clear that in some cases, but not others, existing authorities have to be respected. For example, you cannot understand law without looking to see what the law courts have already decided, as the very meaning of legal words depends on the history of law. Similarly, the religious leap of faith is to accept the word of prophets. However, Pascal was also capable of penetrating scepticism, and he was very clear that when it comes to scientific knowledge, a public experiment carries more weight than the word of Aristotle himself. Indeed, his public demonstrations of the behaviour of barometers were a real turning point in establishing experiment as the ultimate scientific authority.

Pascal's genius was evident from an early age, and when he was only nineteen, he realized that his father's job of calculating taxes could be done by a clock-like device. For example, you can represent the procedure of increasing an integer by one by rotating a cog by one notch. Pascal was inspired to construct a device that physically embodied the basic laws of taxation, and he spent three years refining one of the world's first mechanical calculators. These ingenious devices were known as 'pascalines', and although a few curious individuals purchased them, the idea was literally 300 years ahead of its time, so in 1652 the production of pascalines was halted.

Pascal also deserves to be celebrated as a founding father of statistics: a discipline that lies at the heart of all of modern science. Statistics is the art of reasoning from incomplete information, and more than anyone who lived before him, Pascal pondered the relationship between reason and uncertainty. He helped to invent the mathematics of probability, and he was also one of the key figures in developing the modern understanding of air pressure, vacuums, and other related phenomena. Not only was Pascal a first-rate scientist and mathematician, he was also the author of one the most vivid portraits of the human condition ever written, the *Pensées*, which defends Christianity by examining the roles of reason, emotion, faith and uncertainty in the life of man. And if his many intellectual achievements were not enough to win our admiration, he was also a remarkably charitable man. Indeed, when he decided to give away most of his possessions, he spent some time thinking about how his wealth could do the most good, which led him to invent the idea of subsidized public transport. Pascal's proposal was to charge people a small sum for catching a specially purchased coach, and all the profits from this scheme were used to relieve the worst effects of poverty.

Reasoning by Recurrence

As well as grounding their work by the analysis of concrete cases, mathematicians have always searched for the most general of truths. It is this impulse that drives mathematics to ever increasing levels of abstraction. There are various ways to establish general results. For example, the introduction to this book contains an informal proof of Pythagoras' Theorem, where we saw that by rearranging four right-angled triangles we can turn an 'a-square' and a 'b-square' into a 'c-square'. Because the argument refers only to general properties of right-angled triangles, and we don't need to know the actual values of the lengths a, b and c, we can correctly conclude that Pythagoras' Theorem is true for every right-angled triangle, and not just the particular one shown in our drawing.

There is another fundamental method for proving general statements, which was first explicitly stated in Pascal's *The Arithmetic Triangle* (1654). This technique is rather misleadingly known as 'proof by induction'. To gain some understanding of this general form of proof, let's start by considering the following demonstration that:

$$1+2+3+4+5=\frac{5^2+5}{2}:$$

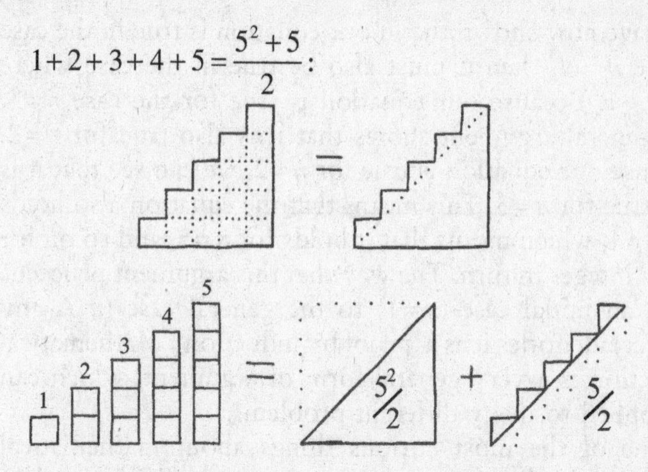

It should be clear that the same area is measured whether we break our shape into component squares and count them, or component triangles. This means that our equation must be true, but what about the more general formula $1+2+3+\ldots+n=\frac{n^2+n}{2}$? How can we be certain that this statement is always true? In other words, how can we be sure that our equation is correct for every integer n, and not just for $n=5$? Unlike our drawing for Pythagoras' Theorem, we cannot simply say that our integer n is irrelevant, as salient features of the drawing depend on this particular value. Despite this fact, it is still possible to prove the general case.

When $n=1$, our general equation reduces to the specific case $1=\frac{1^2+1}{2}$, which is certainly true. Furthermore, we can prove that if our equation is true in the case where $n=N$, then it must also be true in the case where $n=N+1$. That is to say, if we assume that $1+2+\ldots+N=\frac{N^2+N}{2}$, then it follows that:

$$1+2+\ldots+N+(N+1)=\frac{N^2+N}{2}+(N+1), \text{ and}$$

$$\frac{N^2+N}{2}+(N+1)=\frac{N^2+2N+2+N}{2}=\frac{(N+1)^2+(N+1)}{2}.$$

We have now shown that if our equation is true in the case where $n=N$, then it must also be true in the case where $n=N+1$. Because our equation is true for the case $n=1$, our general argument shows that it is also true for $n=2$. Because our equation is true for $n=2$, we can see that it is also true for $n=3$. This means that the equation also holds for $n=4$, which means that it holds for $n=5$, and so on for every integer in turn. The way that this argument proceeds from an initial case ($n=1$) to the general case (n is any integer) identifies it as a 'proof by induction'. Mathematical induction is a very general form of argument, which can be applied to many different problems.

One of the most curious things about mathematical induction is that although it was first explicitly used as recently as 1654, the concept has a very long prehistory. For example, Euclid's proof that there are infinitely many primes implicitly involves something very similar to the modern form of proof by induction, but the fundamental 'inductive' principle was not explicitly identified. Indeed, I suspect that the oldest form of argument closely related to proof by induction is genuinely prehistoric, as we can imagine someone in the distant past making the foolishly claim that 'I have counted as high as can ever be counted'. We could show that they are mistaken by counting one higher, perhaps adding one mark to their tally. If they then said, 'OK, I have counted one higher, and by doing so I have reached the highest number that can be counted,' we should protest at their stupidity. After all, our ability to add an extra mark to their tally does not depend on the number of marks that are already present!

In other words, our objection can always be repeated, and so there can be no integer that is larger than all the others. A similar sense of conceptual necessity is at play when we construct a proof by induction. We start by showing that something holds in the particular case where $n=1$, then we use the general or inductive argument to prove that this implies the result for $n=2$, and because we accept that this second step of the argument can always be repeated, we are logically compelled to conclude that the result holds for the entire, infinite sequence of the natural numbers.

Although we have long appreciated the possibility of endlessly repeating steps in our reasoning, Pascal deserves most of the credit for explicitly developing the two-stepped method of inductive proof that we are taught today, and he used it to demonstrate some enormously significant algebraic results. Another notable moment in the history of mathematical induction came in 1713, when Jacques Bernoulli used an inductive proof to bring some much needed rigour to the rapidly developing field of mathematical analysis. It may have been slow in coming to the conscious forefront of mathematical thought, but by the latter part of the eighteenth century induction was a widely recognized tool in Europe's mathematical armoury.

The Mathematics of the Infinitely Large

Although the infinite was an important and contentious concept for the Greeks, it never occurred to them to compare the infinity of integers with the infinity of points on a line. The idea that there is more than one kind of actually infinite number was first articulated in the second and third centuries BC, by Jaina mathematicians working in India. Jains believe that contemplating very large or infinite numbers has a mind-expanding spiritual value, and Jain cosmology was clearly a crucial factor in stimulating discussion of the mathematically infinite. According to G. G. Joseph's *The Crest of the Peacock*, Jain mathematicians recognized five distinct kinds of infinite numbers: 'Infinite in one direction, infinite in two directions, infinite in area, infinite everywhere and perpetually infinite.'

It is hard to be certain, but the mathematical traditions of Jainism probably didn't impact on the European understanding of the mathematically infinite. There is a strong argument for claiming that the crucial factor re-igniting the question of infinity was the long, complex but ultimately decisive shift from the study of numbers to the study of number systems. Ancient peoples studied numbers and the properties we can demonstrably ascribe to them, but in the nineteenth century mathematicians began to compare systematically entire number systems. For example, it is a characteristically modern observation to note that the integers and the real numbers are structurally similar, in that adding or multiplying an integer by an integer yields an integer, while adding or multiplying a real by a real yields a real.

As well as demanding a level of formal precision and rigour that was often lacking in earlier work, nineteenth-century mathematicians distinguished themselves by studying an ever-expanding and exotic range of mathematical objects. This proliferation of abstract objects put new pressure on the foundations of mathematics. The result was a remarkable growth in the field of mathematical logic (where fundamental work is done to this day), and the development of set theory.

It was in this context of radical innovation that George Cantor (1845–1918) began to rethink the way we approach the infinite, constructing arguments and establishing proofs that directly referred to actual, infinite sets. Previous mathematicians had said that there are infinitely many integers, but they did not think of there being an actual number corresponding to the total number of integers, precisely because no finite number could be big enough. Cantor's controversial claim was that there is a kind of number which is equal to the total number of integers, but this number is a transfinite number, and not a finite integer.

Cantor's Pairs

George Cantor spent his early childhood in St Petersburg, but his family moved to Germany when he was eleven.

As a student he worked in some of the finest mathematics departments in the world, but unfortunately his career as a professor was much less successful than it could have been. Stuck in a minor university, Cantor is now famous for a series of six articles that were published between 1879 and 1884. These remarkable papers formed an introduction to the radical new kind of set theory.

Cantor was gripped by the philosophical implications of his work, and as a devout Lutheran, he believed that his work on transfinite numbers was directly connected to the mind of God. Cantor's work was controversial, and was bitterly opposed by many academics, including the great Poincaré. The mathematician Leopold Kronecker (1823–1891) was a particularly harsh critic, dismissing Cantor's work as 'humbug' and 'mathematical insanity'. Like many people before him, Kronecker believed that mathematics' soundest and most fundamental base must be a number theory derived from counting, as he thought that the only legitimate mathematical concepts were those that can be 'constructed' in a finite number of steps. Kronecker was even suspicious of certain Euclidean proofs, and his attitude can be summarized by his famous remark, 'God made the integers; everything else is the work of man.'

Cantor suffered from depression, and was understandably distressed by the vehement opposition that his work encountered. He desperately wanted to teach in Berlin, but Kronecker could block his appointment, and as the editor of one of the two major mathematics journals, Kronecker could also limit his opportunities to publish. Cantor's radical work on set theory and transfinite mathematics was ultimately influential, but his ideas did not gain widespread acceptance until the beginning of the twentieth century. It is worth emphasizing that Cantor did not win the argument on philosophical grounds (where the debate, though shifted, continues to this day). His principles were ultimately accepted because the set-theoretic framework that his work inspired is incredibly useful. As David Hilbert and many other mathematicians realized, almost all the different branches of mathematics can be described in set-theoretic terms, including the disciplines that were new at the time, such as topology or real function theory.

Two of Cantor's mathematical ideas are exceptionally simple and profound. The first of these truly great ideas was developed with one of the greatest logicians of all time: Gottlob Frege (1848–1925). Essentially, the two men constructed a mathematically fruitful definition for when two sets contain the same number of elements, and they did this by recognizing the deep significance of the following kind of everyday situation. Imagine getting on a bus where every seat has a different person in it, and where there are some people without seats. This observation tells us that there are more people than seats *without counting either number*. Cantor built on this by recognizing the fundamental role that can be played by a one-to-one mapping between sets. In our bus example we have a one-to-one mapping from the set S (the seats) to the set P (the people), precisely because every seat has a different person sitting in it. To put it another way, we can make a rule such that when we are given a seat as input, we return the person sitting in it as an output. This rule is known as a 'one-to-one mapping' because every different input results in a different output.

Every seat can be associated with a unique person (namely the person sitting in that seat), but conversely it is impossible to give a one-to-one mapping from P to S. We cannot allocate every person a different seat when there are more people than seats! We can generalize this principle by saying that a set B is at least as big as a set A if and only if there is a one-to-one mapping from A to B. In other words, if every element of A can be paired up with a different element of B, then we say that B is at least as big as A. Similarly, A and B are said to be the same size if and only if A is at least as big as B, and B is at least as big as A. It turns out that this definition is equivalent to saying that two sets are the same size if and only if we can pair up all of their respective elements.

This is a natural extension of the everyday notion of size – if we have seven apples and seven oranges then we can match them up one against another, and this observation is part of the meaning of 'seven'. However, because this notion of size does not involve counting we can also apply it when reasoning with infinite sets. That is to say, infinite sets can also be paired up. For example, every integer n can be paired with a unique even number $2n$, or a unique square number n^2.

The possibility of pairing up the integers with the square numbers implies a truth that Galileo stated, long before the work of Cantor: 'Neither is the number of squares less than the totality of numbers, nor the latter greater than the former.' Also notice that we can set about generating a list of every positive integer. Similarly, we can generate a list of every (positive) even number. Since every element appears somewhere in the defining list, we say that these sets of numbers are countably infinite. Because we can pair elements according to place (the first thing in list one is paired with the first thing in list two, and so on), we can correctly conclude that every countably infinite set is the same size.

Many people are somewhat perturbed to find that a set can be the same size as one of its subsets. For example, there are as many integers as there are even numbers, even though half the integers are odd. The expectation that a subset must be smaller comes from our experience of counting the finite, but these expectations do not apply at the infinite level. After all, half an infinite set is still infinite! So what about the set of all fractions? How does that compare in size to the set of all the integers? If you try to count the fractions on a number line by working from the smaller fractions to the larger ones, it should be clear that you have to miss most of them out. Whichever fraction you count first will be larger than infinitely many other fractions, and more generally every gap between successive fractions on your list will contain infinitely many rational numbers. Despite this fact, it is actually possible to specify a list that contains every single fraction. In other words, the rational numbers are countably infinite.

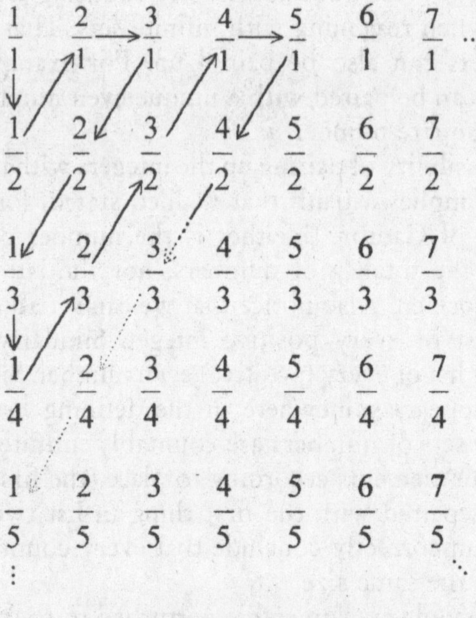

$$\begin{array}{ccccccccc}
\frac{1}{1} & \frac{2}{1} & \frac{3}{1} & \frac{4}{1} & \frac{5}{1} & \frac{6}{1} & \frac{7}{1} & \cdots \\
\frac{1}{2} & \frac{2}{2} & \frac{3}{2} & \frac{4}{2} & \frac{5}{2} & \frac{6}{2} & \frac{7}{2} & \\
\frac{1}{3} & \frac{2}{3} & \frac{3}{3} & \frac{4}{3} & \frac{5}{3} & \frac{6}{3} & \frac{7}{3} & \\
\frac{1}{4} & \frac{2}{4} & \frac{3}{4} & \frac{4}{4} & \frac{5}{4} & \frac{6}{4} & \frac{7}{4} & \\
\frac{1}{5} & \frac{2}{5} & \frac{3}{5} & \frac{4}{5} & \frac{5}{5} & \frac{6}{5} & \frac{7}{5} & \\
\vdots & & & & & & & \ddots
\end{array}$$

As we zigzag through the diagonals of this grid, we eventually hit every positive fraction. We can list every single fraction by starting with 0, and then working our way through the grid shown above, repeating every element in both positive and negative form.

A similar image shows that if we have countably many countable lists, we can combine them into a single countable list. All we need to do is replace row one with our first list, row two with our second list, and so on. Our zigzag path along the diagonals then incorporates every element from every list. Another fact that many people find surprising is that the number of (real) points on a line is the same as the number of points on a square. To see that this is so, recall that each point on the number line can be identified by its distance $0.a_1a_2a_3a_4\ldots$ from the origin. According to Dedekind's definition of a line, a line is just the set of all such points, and given a particular point (i.e., given a particular sequence of digits $a_1, a_2, \ldots$), we can pair the point in question with a point inside a square, namely the point with coordinates ($0.a_1a_3a_5 \ldots$, $0.a_2a_4a_6 \ldots$). Because the points on a line can be paired with the points in a square, it follows that there are the same number of points on a line as the number of points in a square: an infinite number that is provably larger than the number of integers.

The Diagonal Argument

At this point you would be forgiven for thinking that every infinite set is the same size. To see that this is not the case, we must consider the following question: 'Is it possible to list every real number in turn?' Cantor's way of answering this question was the second of his exceptionally deep but simple ideas. His argument constitutes a novel kind of proof, the basic form of which is frequently borrowed by mathematicians. Perhaps the easiest way to explain his argument is to tell a story, where we imagine that George Cantor has been challenged to a guessing game by Lister the Number Genie:

Lister: George, I am challenging you to a guessing game. I have a list of real numbers, represented by their decimal expansions. If you can guess a real number that isn't on my list, I will grant you a wish.

George: Very well, I accept. Can I see this remarkable list of yours?

Lister: Of course, here it is:
0.12345…
0.18345…
0.67391…
0.23475…

George: This is ridiculous. Do you seriously expect me to list an infinitely long real number?

Lister: No, no. I haven't got all day you know. Just tell me your general method for calculating the *n*'th digit, and I will do the rest.

George: OK. The first digit of the first number on your list is 1, so to ensure that my number is different from the first number on your list, I am picking 3 as my first digit. The second digit of your second number is 8 (i.e. not 3 again), and so the second digit of my number is another 3. The third digit of the third number on your list is 3, so to ensure that my number is different from the third number on your list, I shall pick a number whose third digit is 2.

Lister: You're making your number up as you go along, you filthy cheat!

George: Are you sure it isn't you that's cheating? I mean, you do have a definite fixed list don't you?

Lister: Of course.

George: Well then, I have a definite fixed number, namely the number you get by going down the diagonal of your list swapping 2s for 3s, and replacing every other digit with a 3.

Lister: OK, OK – I give in. Your 'diagonal number' must differ in at least one decimal place to every number on my list. For example, it cannot equal the millionth number on my list, because by definition the millionth digit of your number is different from the millionth digit of the millionth number on my list. I don't suppose you fancy a rematch?

George: Will I get another wish if I win?

Lister: Sure, but this time you have to provide the list.

George: Fair enough. Is it all right if instead of comparing digits, we compare finite definitions of numbers, written out in ordinary mathematical language?

Lister: That's fine by me. Now tell me, what kind of list are you going to use?

George: Alphabetical, and by order of length.

The set of definable numbers is countable, because every collection of names or definitions can be listed systematically. You simply start with the shortest names or definitions and move on to the longer ones, listing them alphabetically. Although the set of definable real numbers must be countable, Cantor's diagonal argument proves that every list of real numbers is necessarily incomplete. He showed that given any list whatsoever, we can define a number that isn't on the given list. We simply define a number that is different to the first number in our list because it has a different first digit. Similarly, by definition our number must be different to the second number on the list because it has a different second digit, and so on. On the other hand, the set of finite definitions can be put into an alphabetical list. This argument shows that the set of real numbers must be uncountable, and most real numbers cannot have a finite definition.

Perhaps the most important point is that Cantor exhibited two clear sizes of infinity. So, for example, the set of all integers is countably infinite, while the set of all real numbers is uncountably infinite. This hierarchy of size can also be extended in a natural fashion to contain infinitely many different sizes of infinity. More specifically, Cantor showed that given any set, the set of all of its subsets must be larger than the original set. This means we can construct an infinite sequence of larger and larger infinite sets: first we have the set of all integers (say), then we have the set of all subsets of the integers, then we have the set of all subsets of the set of all subsets of the integers, and so on.

The other critical point is that because the idea of a set is so very general, a vast range of mathematical structures can be described in the language of sets. Partly for this reason, the notion of a set has become central concept of mathematical logic. The task of a logician is to think about thinking. Thinking about thinking is very difficult, but some of the most important intellectual developments in human history have emerged from our attempts to characterize the full extent of legitimate mathematical reasoning. As we shall see in the following chapter, the first few decades of the twentieth century witnessed an explosion in mathematical logic, as many of the greatest mathematicians strove to describe 'the foundations of mathematics'. In particular, the great logician Gottlob Frege helped to bring set-theoretic concepts to the forefront of the mathematical discourse, using the formalism of set-theory to elucidate essential features of mathematical reasoning itself.

Chapter 9:
THE STRUCTURES OF LOGICAL FORM

'Really good systems of logic, says Alembert, are of use only to those who can do without them. Through a telescope the blind see nothing.'

Georg Christoph Lichtenberg, 1742–1799

The Formal Logic of AND, OR and NOT

The study of logic is an ancient pursuit, but it is fraught with the most serious of difficulties. The principle barrier (or, rather, the unnerving lack of barriers) is the nature of the subject matter itself. For example, a good legal argument must be 'logical', avoiding self-contradiction and contradiction with established law, but the skill of developing such a case is part of the art of law itself. Similarly, a good scientific argument needs to be logical, but the skill involved is science itself. The difficulty for the logician is that they aim to study the basis or framework for our valid deductions, but they are interested in doing this for the most general or abstract of contexts imaginable, where we literally do not know what subject is under discussion.

It is difficult to consider logical deduction regardless of the subject at hand, but that is precisely what logicians have tried to do. As Aristotle observed in the *Organon*, there are certain kinds of deductive pattern that crop up time and time again. The basic metaphor underpinning this form of deduction is that the things of this world can be put into categories, and it seems that we think about categories as though they were spatial containers. Cognitive scientists believe we have an innate capacity to appreciate 'container schemas', such as the example of a jar inside a fridge that we considered in Chapter 1. That is because people who have never even heard of mathematics can deduce that if a jar is inside a fridge, and an olive is inside the jar, then the olive must be inside the fridge.

Since at least the time of Aristotle, we have understood that the properties of a thing can place that thing in a certain category (namely, the category of things with the given property), and some categories are understood as being contained within another, larger category. For example, we might accept that the category of 'men' is contained within the larger category of 'mortal things'. It follows that if Socrates is in the category of men, Socrates must also be in the category of mortal things, just like an olive that is inside the jar must also be inside the fridge.

Crucially, this example of deduction fits a general pattern, and for millennia Aristotle's account of proper deduction was considered to be synonymous with valid, logical thinking. Logical deductions lead us from one statement to another, so every deduction begins with an initial premise. For example, we might begin by accepting that the jar really is contained within the fridge, or that the category of men really is contained within the category of mortal beings. Of course, in daily life we don't always explicitly refer to an image of one category contained within another, or worry too much about what kinds of things our categories contain. For example, consider the following statement: 'If it is snowing, then it must be cold outside.' We might accept that this statement is true, and prefer this phrasing to saying that the category of 'occasions when it is snowing' is contained within the larger category of 'occasions when it is cold'. After all, we aren't really sure where the category of 'occasions when it is cold' is meant to begin or end, but in a sense that doesn't matter. The essential point is that if I accept the statement 'If it is snowing then it must be cold outside,' and I also agree that it is snowing, I would be crazy to think that it is warm outside.

We understand that if I say that 'It is snowing but it is not cold outside', I am contradicting my previously stated belief that 'If it is snowing, then it must be cold outside.' Furthermore, it is the *general form* of the 'if ... then ...' assertion that enables us to make this deduction. Indeed, that is the crucial point: to make the deduction, we do not need to interrogate the meaning of the words 'snows' and 'cold', we simply need to understand the logical significance of 'if ... then ...'.

We can demonstrate this point with a second example. If I were to say, 'If Gobbledygook, then Flibbertigibbet,' we can sensibly agree that the logical significance of my 'if ... then' assertion is that whenever I accept the statement 'Gobbledygook', I must also accept the conclusion 'Flibbertigibbet', precisely because I have already agreed that 'if Gobbledygook, then Flibbertigibbet'. In other words, the following is a legitimate deduction, where the conclusion (written below the line) is logically entailed by the premises (written above the line):

'If Gobbledygook, then Flibbertigibbet.'
'Gobbledygook.'

Therefore 'Flibbertigibbet'.

A few decades after Aristotle's death, Euclid brilliantly demonstrated the power of orderly, deductive reasoning. His axiomatic approach showed that the large and complex catalogue of geometric truths known in Ancient Greece could all be deduced from a remarkably simple collection of fundamental axioms. Many of Euclid's deductions subtly relied on geometric intuition, but deductions of the kind demonstrated above can be carried out in an automatic, 'mindless' fashion, without the need to picture the subject matter at hand.

Indeed, we can make valid deductions simply by specifying the proper way to use logical words such as 'AND', 'OR' and 'NOT'. So, for example, we can make the working assumption that our 'atomic statements' A and B are either 'true' or 'false'. Now, by definition, we say that the statement 'A AND B' is true if and only if the statements 'A' and 'B' are both true. That might sound like a circular definition ('AND' means 'and'), but the crucial point is that we can use logical words to generate sentences from other, smaller sentences. For example, given the statements 'A', 'B' and 'C', we can construct the statements 'A AND B', '(A AND B) OR C', 'NOT ((A AND B) OR C)', and so on. Furthermore, the 'truth' of each of these compound statements is entirely determined by the 'truth' of their component pieces. In other words, we can draw up definitive tables that tell us when we should assent to a compound statement given nothing more than the truth or falsity of the atomic statements 'A', 'B' and 'C'.

Many of the statements that we can construct in this fashion are said to be logically equivalent. We say that two sentences are logically equivalent if they are true or false in exactly the same cases. For example, 'A AND A' is true if and only if 'A' is true, so we say that these sentences are logically equivalent. Similarly, in classical logic the statement 'A' is logically equivalent to 'NOT (NOT A)'.

The assertion 'if A is true, then B is true' is particularly important, and in formal logic this is typically written '$A \mapsto B$'. This statement asserts that whenever 'A' is true, 'B' is also true. It follows that '$A \mapsto B$' is false if and only if 'A' is true but 'B' is false. That is to say, there is only one state of affairs that contradicts the claim that 'A↦B', namely the case where 'A' is true but 'B' is false. In other words, in classical logic the statement 'A↦B' can be rewritten in the logically equivalent form 'NOT (A AND (NOT B))'.

Classical Logic and the Excluded Middle

The statement 'A OR (NOT A)' can be represented by the following diagram:

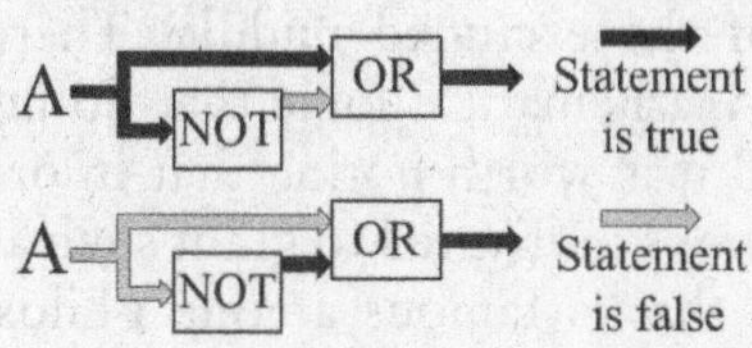

Our scheme presents two acceptable possibilities: the case where 'A' is true, and the case where 'NOT A' is true. In either case, 'A OR (NOT A)' must be true, by definition of the word OR. Statements that are true in every case are called 'logical truths' or 'tautologies'. Because it does not matter what the statement 'A' is, we can plug any formal sentence into the form 'A OR (NOT A)', and get another logical truth. For example,

'((A AND B) OR C)' OR '(NOT ((A AND B) OR C))'

is also a logical truth.

The crucial point to understand is that logical truth is a testable property. That is to say, given any logical structure (constructed by using the words AND, OR and NOT), we can plug in a 'valuation' of the form 'A is true, B is false, C is true, etc.', and systematically work our way along the sentence in question. We eventually either get 'true' or 'false' as an output. Since there are finitely many possible valuations, we can test and see if any given statement is a logical truth. That is to say, given any such sentence, we systematically check to see whether or not every possible valuation gives the output 'true'.

When we are using a formal language, it really doesn't matter what our atoms 'A' and 'B' assert, though it is very significant that in every case we assume that our atoms must be unequivocally 'true', or unequivocally 'false'. This deep-rooted feature of classical logic is known as the law of the excluded middle. There are modern variations of mathematical logic that do not assume this principle, and it is worth noting that in ordinary speech we are not always motivated to grant such an assumption. The example that is famous among philosophers is the statement 'Hamlet's grandmother has blue eyes'. Since we have no reason to assert that there are any facts whatsoever concerning Hamlet's grandmother, it may seem inappropriate to think that this statement must be either 'true' or 'false'. However, even though Hamlet's grandmother is an utter phantom (being an as-yet-unwritten fictional character), we must admit that it is quite impossible to cast a film tracing Hamlet's ancestors without coming to a decision as to whether or not she has blue eyes.

Ordinarily we distinguish between factual accounts and works of fiction by engaging with the wider world, looking for the occasion to which our words might fit. In the peculiar case of mathematics there is clearly a distinction between the factual and the false, but the discipline is unique in that mathematicians have no grounds for sustaining a distinction between the factual and the fictional. We cannot make such a distinction because the thing in the world that testifies to mathematical truth is the mathematical utterances themselves. In other words, mathematical language brings into being the very truth that it proclaims.

Mechanical Deductions

If we are given a sentence constructed out of the words AND, OR and NOT, we can systematically check to see whether or not our statement is true in each and every case. So, for example, we can see that 'A OR NOT A' is true in every case; 'A OR B' is sometimes true; while 'A AND NOT A' is always false.

Furthermore, we can be very efficient in the way that we identify logical truths. It is possible to make a machine (e.g. a computer) that uses only a few different rules to generate every logical truth in turn. The basic idea is that every finite collection of statements can be used to generate a finite number of 'logical offspring'. For example, the statements 'A' and 'B' can be used to produce the logical offspring 'A AND B'. Similarly, the statements 'A OR B' and 'NOT A' can be used to produce the logical offspring 'B'. Notice that in the first example we have two shorter sentences producing a longer one, while in the second we have two longer sentences producing a shorter one.

The essential point is that given any collection of statements we can systematically set about generating their logical offspring. For example, we can feed a finite list of statements into a computer, and mechanically generate a further list of statements, which I have described as the logical offspring of the original statements. If we start with logical truths then we generate offspring that are also logical truths. If we start with a statement such as 'A' (which is not a logical truth), we will generate more than just the logical truths. In fact, the statements that we generate are precisely those statements that are true whenever statement 'A' is true (i.e. for every valuation that says 'A is true'.) For example, given the statement 'A', our machine will eventually generate the statements

'A OR B' and 'A AND (B OR (NOT B))',

because both these statements must be true whenever 'A' is true. The inputs for our logic machine are called axioms, while the statements our machine generates (given an input A) are called 'the logical consequences of A'. Arguments in the language of AND, OR and NOT (or AON for short) are usually called 'derivations'. These are just chains of statements, starting with the axioms, such that every statement is the offspring of the statements that precede it.

Arguments of this form show that if we believe the axioms, then we should also believe every sentence in the argument, precisely because we have accepted practice for using the words 'AND', 'OR' and 'NOT'. Indeed, this is the heart of logical analysis, as we are able to make logical deductions precisely because we use logical words when we state our axioms. That is to say, it is the logical structure of our axioms that enables us to draw conclusions. To put it another way, deduction is very important, but when we make a valid deduction, we cannot learn anything that wasn't already implicit in the axioms themselves.

Quantifiers and Properties

For over two millennia, logic barely progressed beyond the ideas of Aristotle. Two people who did make some progress were Gottfried Leibniz (1646–1716) and George Boole (1815–1864), who had the brilliant idea of replacing the logical terms of ordinary language with symbols and symbolic operations. In other words, Leibniz was prescient enough to dream of turning logical deduction into a form of calculation, while Boole successfully developed a formal language equivalent to AON. This was a remarkable innovation, but there is very little mathematics that can be done if your only logical words are AND, OR and NOT.

The idea that mathematics is or should be an exercise in logic first became truly plausible with the work of

Gottlob Frege (1848–1925). The pivotal moment occurred in 1879, when Frege published the enormously influential *Begriffsschrift* (or 'Concept Script'). This famous work contained an incredibly powerful system of formal logic, which Frege and others developed over the following decades. In essence, he recognized that we can produce a perfectly effective automatic logic machine for formal languages that are much more complex than AON. This is crucial, because to rigorously state their arguments, mathematicians need a formal language that includes the phrases 'EVERY thing' and 'SOME thing'. Without these words we cannot ask difficult questions about the integers. For example, we could not ask whether it is true that 'for EVERY integer n, there is SOME sequence of prime number $p_1, p_2, \ldots, p_m$ such that $n = p_1 \times p_2 \times \ldots \times p_m$'.

One of the first things to note about these logical words is that when we use the words 'EVERY thing' or 'SOME thing', it doesn't really matter what kind of 'thing' we are talking about. This comment may sound rather peculiar, but remember, in formal languages such as AON, it really doesn't matter what our atoms A, B and C refer to. That is to say, we can understand statements in this language as a finite collection of 'meaningless' symbols, because it is the words AND, OR and NOT that do all the work when we make our deductions. Similarly, in Frege's more advanced logical language, it is the words AND, OR, NOT, EVERY and SOME that do all the work, so in a sense we don't need to worry about the meaning of any other term.

As well as having symbols for EVERY and SOME, mathematicians also require symbols that indicate properties, as an integer may or may not be even, square or triangular (say). It is not that each new mathematical concept requires a fundamentally new form of logic, it is just that when we write our mathematical sentences in formal language, we need some kind of symbol to express the idea that a given thing has a given property. As far as the logic is concerned it doesn't really matter what these properties are, but it is a fundamental truth that in this language, EVERY thing either has property P or does NOT have property P. If the property in question is a bit fuzzy, so it does not make sense to assume that things either have the property of not, Frege's logic system is not appropriate, and we need to use a different kind of logical language.

The basic idea is that Frege's logical language contains the words AND, OR, NOT, EVERY and SOME, together with an arbitrary number of 'variables' and 'properties' (or predicates). Not only did Frege provide axiomatic definitions for these basic logical terms, he also constructed a strictly rule-governed, symbolic system for representing logical deductions, rather like a modern computer language. This formal system is called 'predicate calculus', and I shall refer to it as PC for short. Strictly speaking this language doesn't contain the ordinary English words that I have used to make sentences in PC more easily readable. For example, we do not need to bother writing the words 'thing', 'has' or 'property', although some kind of basic grammar is required. Another somewhat technical point is that in the language PC, there is a correct way to use brackets, and every grammatical sentence contains equal numbers of left and right brackets. I am not going to run through the details, but some of the key points of PC are the following:

1. In PC, the statement 'NOT (EVERY thing has property P)' is logically equivalent to the statement 'SOME thing does NOT have property P'. Note that when we are using the language PC, we are not obliged to make any kind of claim about our ability to actually find this thing (whatever it may be).

2. Throughout this book, whenever I am phrasing a sentence in the formal language of PC, I will capitalize the words AND, OR, NOT, EVERY and SOME. This is to remind the reader that the sentence in question can be precisely phrased in terms of a formal language, and it is the capitalized words that are critical when we are making logical deductions.
3. Given a list of grammatically correct sentences 'I', we can imagine a PC machine that sets about generating an infinite list of their logical offspring, called 'the logical consequences of I'. Inputs of the form 'A OR (NOT A)', or 'EVERY thing (either has property P OR does NOT have property P)' are called 'logical inputs', because these statements are true for every possible valuation.

In his doctoral thesis of 1930, Kurt Gödel proved that when we use logical inputs, PC generates every logical truth in the given language. This important fact is known as 'the completeness of predicate calculus'. In effect, Gödel proved that we can systematically generate every statement which is true for every collection of things, and every property P. An example of one of these tautological statements would be: 'It is NOT the case that EVERY thing has property P AND SOME thing does NOT have property P.'

Inputs for Predicate Calculus

Logical inputs are not the only kind – we can also use mathematical axioms as inputs for PC machines. For example, we might use the language of PC to state the following eminently sensible axiom: 'For EVERY integer *x* there is SOME integer *y* such that *y* is bigger than *x*.' This logical statement is what we mean when we say that there are infinitely many integers. Similarly, we can use the language of PC to define more sophisticated mathematical concepts.

For example, in the early nineteenth century (some fifty years before Frege published his works on logic), mathematicians were eager to develop new branches of calculus. In order to extend the fundamental ideas to the most general case, they needed to specify or define what is meant by a 'limit case'. Bernhard Bolzano (1781–1848) and Augustin-Louis Cauchy (1789–1857) independently hit upon the same definition: a sequence $x_1, x_2, \ldots$ 'converges to a limit L' if and only if 'for EVERY positive number δ, there is SOME number n such that EVERY term $x_n, x_{n+1}, \ldots$ is bigger than $L - \delta$ AND smaller than $L + \delta$'.

One cannot overstate the importance of defining mathematical concepts in terms of a logical language, as such an approach leaves us in no doubt about the kinds of deduction that we are entitled to make. As Jaakko Hintikka wrote in *The Principles of Mathematics Revisited*, 'We can make the axioms of a typical mathematical theory say what they say only by using [words such as "AND", "OR", "NOT", "EVERY" and "SOME".] If mathematical propositions were not expressed in terms of logical concepts, it would not be possible to handle their inferential relationships by means of logic.' Of course, mathematics is several millennia older than modern, formal logic. Hintikka's point is not that it is impossible to reason without an explicit formal system, it is just that explicit formal systems are of great benefit because they help to elucidate the logical underpinnings that mathematical statements necessarily possess.

If we adopt a system like PC, we can made valid deductions by shuffling symbols in an entirely mechanical fashion without relying on any kind of understanding of what the symbols refer to. In a sense, routine mathematics does not require knowledge of whatever we are talking about: we can articulate the relevant understanding of our subject matter by formulating axioms, and then apply a system like PC in a mindless, mechanical fashion.

As well as elucidating our basic deductive practices, formal languages like PC are enormously useful, because they enable us to construct testable proofs. That is to say, by translating an argument into PC we can be certain that there are no hidden assumptions, and that our conclusion really is a logical consequence of the axioms. In other words, we can show that one should accept a chain of reasoning from the axioms to the conclusion precisely because of the formal practice associated with the words AND, OR, NOT, EVERY and SOME. Furthermore, the axioms that we use often seem to force themselves upon us as inescapable, or 'self-evidently true'. In such cases our common sense compels us to accept their implications quite literally once and for all.

It can be very difficult to construct a formal proof even if the idea behind it seems relatively clear, but once this has been done we can be absolutely certain that there are no hidden assumptions, and that our conclusion really is a logical consequence of the axioms. It is also worth noting that the process of translating an intuitive insight into a strictly formal proof can be very revealing, but this is not always so. It is certainly possible to check a formal proof but still be confused about the nature of the argument. In many ways the most significant thing for the mathematical community is that formal proof provides a clear and comprehensible criteria for when a piece of work is complete and valid, which means that mathematicians can reach a truly exceptional level of consensus.

Axiomatic Set Theory

Frege's arguments were enormously influential, not least because of their effect on the Vienna Circle, and the philosophy of Wittgenstein. In the world of mathematics,

Frege's work inspired the growth of two distinctively modern branches. The first of these is mathematical logic. Bertrand Russell and Albert North Whitehead wrote a classic example of this kind of work: the three-volume *Principia Mathematica* (1910, 1912 and 1913). In this massive tome, the authors derive familiar mathematical results using a highly formal and rigorous system of logic. For example, they use set theory to confirm that '1 + 1 = 2', and they also prove things like Pythagoras' Theorem, and other more sophisticated results.

Over the next couple of chapters we will return to the subject of mathematical logic, particularly the extraordinary work of Alan Turing and Kurt Gödel. First I want to mention the second branch of mathematics that Frege inspired: axiomatic set theory. This branch of mathematics grew from his work less directly, being shaped by developments elsewhere in the mathematical arena. In particular, the study of sets gained new impetus as a result of George Cantor, and his ideas about the infinite. Initially, mathematicians presumed that the concept of a set was utterly basic, as Cantor once described a set as 'any collection into a whole M of definite, distinct objects (that is, the members of M)'.

As we shall see, this ultimately proved to be an inadequate way of articulating the concept of set. The problem was that mathematical logicians, starting with Frege, were interested in the incredibly general notion of a property, where the basic idea is that we can either ascribe a property to an object, or not ascribe that property to an object. Frege was also concerned with the 'extension' of properties, following the idea that every property determines the set of things that have that property. For example, the property of being a dog can be thought to determine the set of all dogs, while in the realm of mathematical concepts, the property of being a prime number is related to the set of prime numbers. At worst, it was believed that a well-defined property might have an 'empty extension', as, for example, there is not a single number with the property of being an even prime number larger than two.

In 1902 Russell posted a famous letter to Frege, saying: 'I find myself in agreement with you in all essentials. ... There is just one point where I have encountered a difficulty.' He then went on to mention a property that a set might have, namely the property of being a member of itself. For example, there are infinitely many sets that have the property of being infinite, and so we might imagine that the set of infinite sets is a member of itself. Conversely, the set of all dogs is not itself a dog, and so this set is not a member of itself. Russell posed a question that devastated Frege's grandest scheme: given that we can conceive of the property '*x* is not a member of itself', what are we to make of the corresponding set R, which consists of all sets *x* such that *x* is not a member of itself?

The problem is that this so-called set is paradoxical. If R is a member of itself, then by definition it must not be a member of itself. Conversely, if R is not a member of itself, then by definition it is a member of itself! Russell's paradox and other related conundrums provoked consternation among many logicians, and the event is sometimes described as a crisis in mathematics. However, the problem was not with building up collections of familiar mathematical objects, as that process has never led to any kind of paradox. Rather, the problematic assumption that underpins Russell's paradox is the naïve belief in a curious and un-mathematical object known as the 'set of all sets'. That is to say, his paradox arises when we try to divide an un-described totality into two distinct categories: every set with property A, and every set without property A. As Gödel sensibly remarked, 'These contradictions did not appear within mathematics but near its outermost boundary toward philosophy.'

As a result of these debates, the naïve conception of a set was formally refined. In particular, Ernst Zermelo (1871–1953) and Adolf Abraham Fraenkel (1891–1965) developed the modern, 'iterative' conception of sets, which is described by a branch of mathematics known as 'ZF set theory'. The basic idea of ZF set theory is to build up new sets from old. The first axiom simply states that there is a set with no members, and we call this set the empty set. The other axioms of ZF set theory tell us how to produce new sets from old. For example, if A is a set then so is {A}, (i.e. the set whose only member is A). Similarly, the axioms of ZF set theory state that if A and B are both sets, then so is $A \cup B$ (where a set is a member of $A \cup B$ if and only if it is a member of A or a member of B). If we adopt the eminently sensible axioms of ZF set theory, the inherently contradictory 'set of all sets' is recognized as not being a set at all, because it is decisively ruled out by the iterative conception.

Because of this axiomatic approach, set theoretic logic is not the same thing as predicate calculus. However, the incredibly general concept of a property is closely linked to that of a set. Indeed, the most important thing about axiomatic set theory is that it unifies mathematics, as a vast range of mathematical ideas can be rephrased in the language of sets. For example, a line can be defined as being a set of points, and the points where two lines intersect is simply the set of all points which belong to both the set that defines the first line and the set that defines the second line. Similarly, the statement 'the integer x is smaller than y' is equivalent to the statement 'the pair of integers (x, y) have the property $x<y$', which in turn is equivalent to saying that 'the pair of integers (x, y) belong to a particular set, namely the set of pairs of integers where the first integer is smaller than the second'. Of course, to make these statements using nothing but the language of set theory, we need to define the integers within the

language of sets, but that is not difficult to do. We simply identify '0' with the empty set, while '1' is identified with the set that contains the empty set and nothing else, so by this definition '1' has one member. The number '2' is identified with the set that contains two members (the empty set '0' and the set '1'), and so on.

Chapter 10:

ALAN TURING AND THE CONCEPT OF COMPUTATION

'If you have clear concepts, you know how to give instructions.'

Johann Wolfgang von Goethe, 1749–1832

From Mechanical Deductions to Programmable Machines

In the previous chapter we saw how mathematics at the beginning of the twentieth century was characterized by an extreme concern for formal rigour, enabled by significant advances in formal, symbolic logic. Although great progress was made, this era of mathematics was marked by a belief in a pair of philosophical propositions that have ultimately proved to be untenable. First, it was believed that the essence of a subject matter can always be given by a small set of axioms, and that those axioms ought to provide a secure foundation on which the entire subject can rest. As we shall, the truths of arithmetic cannot be reduced to a finite set of axioms, so this idea is definitely mistaken. Second, it was believed that mathematical reasoning is a form of calculation, and that all truths can be calculated using mathematical logic. As I discuss in the final chapter, this belief is not justified by the practices of real mathematicians. Nevertheless, we might consider completely explicit, mechanically reproducible forms of reasoning the ideal to which we ought to aspire.

The relationship between mathematics and logic has been grist in the philosophical mill for many centuries, but the genius of Gottlob Frege, and the possibilities presented by his predicate calculus, led many of the brightest minds into the challenge of 'rebuilding' the foundations of maths. In the beginning of the twentieth century, geometry, number theory and other branches of maths were rephrased into the language of sets, so that the vagueness and ambiguity of natural language could be scrupulously avoided. For example, the massive *Principia Mathematica* uses a few axioms of set theory, together with a few rules of inference, and on this basis the authors derive a substantial proportion of ordinary mathematics.

In many ways, this work was the culmination of an ancient scheme. At the very least, Leibniz had dreamed of 'a general method in which all truths of reason could be reduced to a kind of calculation'. In Leibniz's day there was little progress in formal, mathematical logic, but by 1910 the situation had changed. The many 'truths of reason' contained in the *Principia* can all be reduced to a kind of calculation. More specifically, we can adopt the language of the *Principia*, and generate many true statements, simply by following a few, clearly stated principles. The extreme, systematic rigour of such works was philosophically influential, and by trying to fit different areas of study into a common framework, mathematicians revealed many fundamental connections that had previously been obscured.

In 1928, David Hilbert cut to the heart of the matter by posing a challenge known as Hilbert's decision problem. The underlying idea is that given a formal language, there are certain statements we can make using the symbols of that language. Hilbert's challenge was to find a general, mechanical procedure that takes as input any statement in the given language. That is to say, our procedure ought to begin with a simple string of symbols. To positively solve Hilbert's decision problem, our procedure must result in the output 'true' whenever the statement (string of symbols) is true, and it should result in an output 'false' whenever the statement is false.

This was a provocative challenge, which inspired some brilliant work. Intuitionist mathematicians, such as L. E. J. Brouwer, took issue with certain rules of classical logic, particularly the law of excluded middle. They also cautioned against the prospect of mathematics being dominated by mechanical deductions. After all, people become mathematicians by engaging their imaginations. If we view the mental life of mathematicians as the heart of mathematics, we might follow the intuitionist line, and say that formal notation, and the mechanical application of rules, are 'merely' imperfect ways of communicating the true heart of mathematics.

In contrast, Hilbert and his fellow formalists emphasized the proper use of symbols. They argued that the systematic, rule-governed use of symbols is central to the *meaning* of mathematical practice. There are many subtleties and confusions that fuelled this debate, and I shall return to the delicate question of the meaningfulness of mathematics in the final chapter. At this point, I just want to stress that whatever ideas or images we might personally entertain, the meaning of mathematical symbols ultimately depends on the way that those symbols are actually used. The philosopher Michael Dummett put this point well in *The Philosophical Basis of Intuitionistic Logic*:

> An individual cannot communicate what he cannot be observed to communicate: if an individual associated with a mathematical symbol or formula some mental content, where the association did not lie in the use he made of the symbol or formula, then he could

> not convey that content by means of the symbol or formula, for his audience would be unaware of the association and would have no means of becoming aware of it. To suppose that there is an ingredient of meaning which transcends the use that is made of that which carries the meaning is to suppose that someone might have learned all that is directly taught when the language of a mathematical theory is taught to him, and might then behave in every way like someone who understood the language, and yet not actually understand it, or understands it only incorrectly.

Remarkably, these somewhat esoteric arguments in the philosophy of maths have had a very real and substantial impact on the way we live today. As we shall see in the following chapter, Hilbert's decision problem inspired Alonzo Church (1903–1995) and Alan Turing (1912–1954), which directly led to the invention of computer programs. A few years later, this fundamental concept inspired the invention of real, programmable machines, which were first used to crack the Nazis' secret codes. The importance of the code-breakers of Bletchley Park cannot be overstated, as their work may well have changed the entire course of the war. Of course, Turing did not crack the codes single-handed! Over 10,000 people worked at Bletchley Park, and no matter how intelligent our experts may have been, reading the Nazi codes would have been a practical impossibility without some fairly detailed knowledge of the encryption process.

Fortunately, five weeks before the outbreak of war, the Polish military intelligence service presented a gift to their French and English counterparts. At a top secret meeting in Warsaw, intelligence officers were shown a replica of the German 'Enigma' machine. Polish mathematicians had been working on methods of decryption for several years, and a young Alan Turing quickly mastered their techniques. He played a crucial role in developing the British code-breaking routines, and, in particular, he invented a machine called a 'Bombe'. By the end of 1940 these devices enabled the code-breakers to read all messages sent by the Luftwaffe. Naval codes were harder to break, but thanks to some captured information, naval signals could be decoded for most of the months of war.

Turing made a number of vital contributions to the war effort, and we now know that he helped his old logic teacher Max Newman to develop the world's first programmable, electronic computer (the top-secret 'Colossus' of Bletchley Park). Like most mathematicians, Turing pursued a number of academic interests. In addition to his groundbreaking work on computation, Turing developed some remarkable, mathematical ideas relating to the growth of biological forms, which we will return to in Chapter 12. He was also a superb long-distance runner, and was almost good enough to compete at the Olympic level. Tragically, in 1952, he answered some policemen's questions by telling them he was gay. Turing didn't think there was anything wrong with being gay, but despite having the best character witnesses imaginable, he only escaped prison by agreeing to be injected with female hormones. As a result of this 'treatment' Turing developed breasts, and in 1954 he died after biting a cyanide-coated apple.

Depicting Calculation

The modern, mathematical concept of computability dates back to 1936. That was the year that Alonzo Church developed his 'lambda-calculus', while Alan Turing was devising his own approach to the challenge of finding a mechanical procedure that identifies true statements (Hilbert's decision problem). Intuitively speaking, the notion of a mechanical or automated procedure seems to be fairly clear. In this day and age we are used to the idea of computing machines, and we know that these machines need a 'program'. Indeed, everyone knows that computer programs have significant financial value. The important historical point is that the mathematical concept of a 'program' predates real, programmable machines. To make dramatic progress with Hilbert's decision problem, mathematicians did not need actual computing machines. What was needed was a form of symbolism that could be used to represent programs.

Turing made his great advance by imagining a person 'mindlessly' carrying out a computational task. For example, we might imagine a person adding together the numbers on a list, systematically adding the digits in the unit column, then the digits in the tens column, and so on. Such a person starts their procedure with a finite number of symbols printed on a piece of graph paper, and they can carry out their task even if they only view one digit at a time. We can imagine someone placing a marker on the paper at a predetermined point (the top of the units column). They then proceed in a step-by-step fashion, with the marker moving one square at a time. There are definite rules which the 'player' must follow, and given a particular input, there is always one correct thing to do. This may involve getting more paper (our player is assumed to have an unlimited supply), and it may involve writing some 'working out'. This 'working out' is an activity that takes place on the sheet of graph paper, where it is recorded.

Adding a list of numbers is a specific example of a very general occurrence. The nature of a routine calculation is that we begin with a finite arrangement of symbols (the problem), which are then changed one symbol at a time, until we have another finite arrangement of symbols (the answer). The crucial point is that in such a routine calculation, we must specify the one and only 'correct' thing to do at every step. If we have done that, we say that the person performing the calculation is playing a deterministic language game.

Unfortunately, mathematics itself is often misunderstood as a deterministic language game (which it conclusively cannot be, as we shall see later). In practice, mathematical proofs of any real sophistication cannot simply be checked by computers, so the activity of most mathematicians cannot be summarized in terms of simply describable routines. However, individual mathematical procedures or calculations are language games, as mathematical understanding tends to the point where we can instruct others, and then check in an uncontroversial manner whether or not a well-described technique has been implemented properly.

Deterministic Language Games

Turing invented a standardized formal system for describing deterministic language games. As we shall see, any deterministic language game can be described in this manner, no matter how strange its rules might be. A language game that has been described using Turing's system is usually called a Turing machine. However, I want to emphasize the fact that 'programs' are conceptually quite separate from actual computing machines. Indeed, the designers of modern, digital computers were inspired by their study of mathematical logic: the maths came first, and then the physical machines. For that reason, I will describe the mathematical objects in question as Turing cards, not Turing machines.

Once again, the key idea is that we can use a pack of Turing cards to summarize any fixed procedure that converts an input of symbols into an output of symbols. Perhaps the most important examples of such procedures can be found in science, as in many cases we can describe our scientific theories as a mapping between states of affairs.

Scientists and engineers often use descriptive data and a well-stated theory (e.g. Newton's laws) to produce a further statement: the prediction of the stated theory. For example, we might input a description of the location, speed and mass of a star and planet, and output a prediction for the planet's orbit. We can describe this process as a rule-based mapping from one set of symbols to another, so Turing cards are just the kind of thing we need to summarize our rule-based process.

The assertion that every deterministic procedure can be put on a pack of Turing cards is called 'Church's thesis', and we will examine that important claim in the following section. First let's see just how easy it is to use a pack of Turing cards. As we have seen, a deterministic language begins with a piece of graph paper covered in symbols. We can use any finite alphabet of symbols, and as we only ever look at one symbol at a time, it is helpful to imagine a movable marker that tells us where to look. So what about the pack of Turing cards? What does that look like?

Each card in the pack is labelled with a number, and each card contains a list of instructions: one instruction for each of the different symbols that might be written on the piece of paper. When the game begins, we look at the first symbol on the page (that is, the one with the marker on it), and we turn to the card that is sitting at the top of our pack. We then follow whichever instruction we are supposed to follow, given the symbol that is written on the marked square of our graph paper. Turing cards only ever contain two kinds of instruction. The first kind says: 'Erase the symbol on the graph paper, replace it with the symbol x, and put card number n at the top of your pack.' The second kind of instruction says: 'Move your marker one square to the right (or one square to the left, etc.), and put card number n at the top of your pack.'

The game can come to an end because one of the cards simply says, 'Well done, you have finished.' If the player is instructed to turn to that card, the game does indeed finish. Otherwise the player simply carries on following one instruction after another in a deterministic manner. The crucial point is that at every step in every deterministic language game, the correct instruction to follow depends on the answer to two questions: which card is at the top of the pack, and what symbol is written on the marked square?

A simple example of a deterministic language game takes an input of the following form:

We start with *n* dots, a space, and then *m* dots. We begin this particular game in 'drawing mode' (card number one), which tells the player to move the marker to the right until it hits the space. Once the marker hits the space, the player fills in the space with a dot, and then switches to 'scanning mode' (card number two). This card tells the player to move the marker to the right until it reaches another blank, at which point they are instructed to switch to 'erasing mode' (card number three). This tells the player to move the marker back one square to the left, so it is resting over the right-most dot of all. The player is then instructed to erase this right-most dot, before finally being told that the game has come to an end. What is left on the piece of paper is something of the following form:

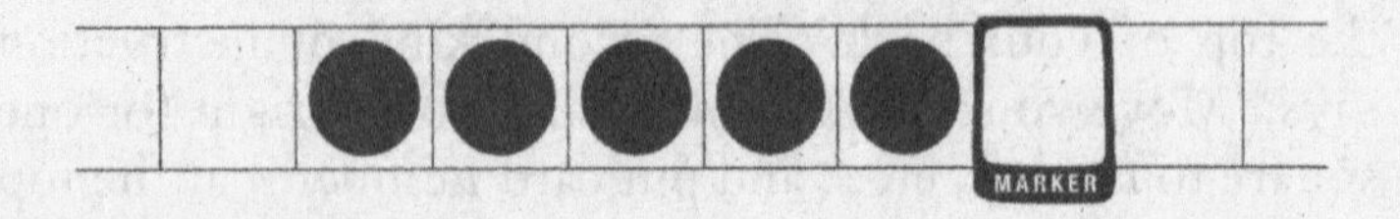

In other words, I have described a simple program for addition, as when we are given n dots and m dots, the game terminates once we have drawn $n+m$ dots.

In a sense, the heart of Turing's argument is that any formal procedure can be described. Such a description is a finite collection of symbols, and any finite collections of symbols can be considered as a mathematical object. We call these abstract objects 'programs'. The crucial point is that by transforming a verb (the procedure) into a noun (the program), we can make the intuitive notion of computability mathematically comprehensible.

Turing's second great insight was to realize that a single pack of cards can enable a person to carry out any computational procedure whatsoever! In other words, there necessarily exists a 'universal computing machine', and such a 'machine' only needs to be about as large and complicated as an ordinary pack of cards. This follows because *it is perfectly possible to describe the proper process for using an arbitrary pack of Turing cards*. The rules that govern the proper use of Turing cards can themselves be summarized into a single, slender pack, which I shall call pack U, for universal.

Now, imagine holding this pack U, which provides a player with all the instructions needed to use any other pack of Turing cards. If you want to carry out any computational procedure whatsoever, all you now require is the appropriate input. For example, suppose that you want to know what happens when a pack T is used, starting with input I. You simply need to copy out some symbols that describe the pack T, together with the input I. The resulting combination of symbols {T, I} now becomes an input for our universal pack U. Crucially, we can be certain that using pack U on the input {T, I} produces exactly the same output as using pack T on input I. Since pack T and input I could summarize any deterministic language game whatsoever, our universal pack U really can carry out any computation we choose. Of course, using a universal pack U might be an absurdly *inefficient* way to carry out the calculation, but if we were given an endless supply of time and paper, we would eventually write out the correct, final answer.

Church's Thesis

The claim that every well-defined, deterministic procedure can be summarized on a pack of Turing cards is not a strictly provable claim. The intuitive concept 'mathematical procedure' is not itself mathematically defined, in contrast to the set of all Turing packs. Despite this observation, we have a very good reason for accepting Alonzo Church's thesis. Suppose that we have some kind of technique T, which generates statements from other statements. For example, imagine using Pythagoras' Theorem to calculate the size of the hypotenuse, given the lengths of the other two sides as an input. There are two ways that we might check that our procedure has been carried out correctly:

1. There is an algorithm for calculating the output, which our technique T correctly follows.
2. There is a theorem that proves that input I corresponds to output O. In this case we can prove that the given relationship between input I and output O follows as a logical consequence of our theorem T.

Both of these things are step-by-step procedures, in which we look at the input, write out working and look over what we have already written (accessing information). But these are precisely the activities that Turing cards are designed to summarize!

In conclusion, if we follow the instructions on a pack of Turing cards, and if we have an unlimited supply of time and paper, we can (in principle) calculate the output of any deterministic algorithmic whatsoever. Turing cards can even carry out information processing procedures where the program itself changes over time. On the other hand, it is worth noting that modern computers often perform tasks by responding to inputs while they are already engaged in a computational procedure, and that kind of interactive process isn't well captured by the formalism I have described.

Theoretical computers have many uses, and in particular, we can use Turing cards to make decisions of a certain kind. For example, a computer can effectively decide whether or not a given integer is prime, and we can use Turing cards to do the same thing. This idea is very important, as we can use it to elucidate the subtle relationship between mathematical truth and mechanical procedures. That fundamental relationship is the focus of the next chapter, but before we investigate these fascinating questions, we must first examine the concept of a 'decision problem'.

Decision Problems

In mathematics, we often say that one set of objects or symbols have a given property, while another set of objects or symbols do not. For example, we can divide the integers into two sets: the numbers that are even, and the numbers that are not. As we can represent the integers using ordinary digits, we can also divide strings of digits into a pair of sets: the strings of digits that represent even numbers, and the strings of digits that do not.

If some things have a property P while other things do not, it is natural to wonder how we might set about answering the question, 'Does this particular thing have property P?' Questions of this form are known as decision problems. If there is a general, mechanical method that provides an answer in every case, it must be possible to summarize that method using a pack of Turing cards. When using Turing cards to tackle a decision problem, we make the safe assumption that our pack contains two special, 'halting' cards. One of these cards says: 'Halt – the input has property P.' The other halting card says: 'Halt – the input does NOT have property P.' None of the other cards in the pack will tell us to stop the computation, though they may instruct us to turn to one of the halting cards.

If a pack of Turing cards can correctly decide whether or not a string of symbols has a certain property P, we say that the problem is 'decidable'. For example, the question 'Is this integer prime?' is a decidable problem, because we only need finitely many axioms and finitely many rules of deduction to determine whether or not a string of digits represents a prime number. Because we can settle this problem using a well-characterized finite system, there must be a pack of Turing cards that can correctly decide whether or not any given integer is prime. On the other hand, if a problem cannot possibly be solved using a pack of Turing cards, we say that the problem is 'undecidable'. The first and one of the most famous undecidable problems is the halting problem. This problem was posed and analyzed by Alan Turing in his classic paper 'On Computable Numbers, with an Application to the Entscheidungs problem'. He asked, 'If I use this pack of Turing cards on that input, will my computation ever come to halt?'

If the combination does result in a finite computation, we can definitely find this fact out. That is to say, there is a well-defined test for every thing of this type, and every thing of this type passes the appropriate test. In fact, the test is very simple: we simply use the Turing cards on the given input, and then check that we are eventually instructed to halt. On the other hand, some programs cannot be efficiently predicted. That is to say, in some cases we can never know that a given computational procedure will never come to a halt.

Turing proved that there is no general method for recognizing non-halting combinations by considering 'self reflective' language games. To understand the proof, the first observation that we need to make is that given a pack of Turing cards T, we can consider what happens when we take, as an input, a standardized description of the pack T. A language game like this proceeds much like any other language game, as we use the instructions in pack T to change the symbols of our description of T. This process results in our description changing step-by-step into another collection of symbols. Some packs produce self-reflective language games that come to a halt, while other packs produce a self-reflective language game that never halts.

Now suppose that we have a pack of Turing cards that correctly determine whether or not a given input has some property P. Recall that such a pack contains two halting cards: one that says 'Halt – the input has property P' and another that says 'Halt – the input does NOT have property P'. Given such a pack, we can construct a modified version by removing the second halting card, and replacing it with a 'dead-end card', which simply tells us to repeat the same, pointless operation over and over again, without ever moving a different card to the top of the pack. If we use this modified pack there is only one way that the computation can ever halt. If the input has property P, we will eventually reach the card that says 'Halt – the input has property P'. As we removed the other halting card, this is the only way that our computation can ever halt.

We are now ready to hear the final part of Turing's analysis. For the sake of argument, he supposed that there is a pack of Turing cards that can recognize when any given Turing pack T produces a non-halting, self-reflecting game. Now, if there was a pack of Turing cards R that can recognize non-halting, self-reflecting games, we would find ourselves in the following situation for every Turing pack T:

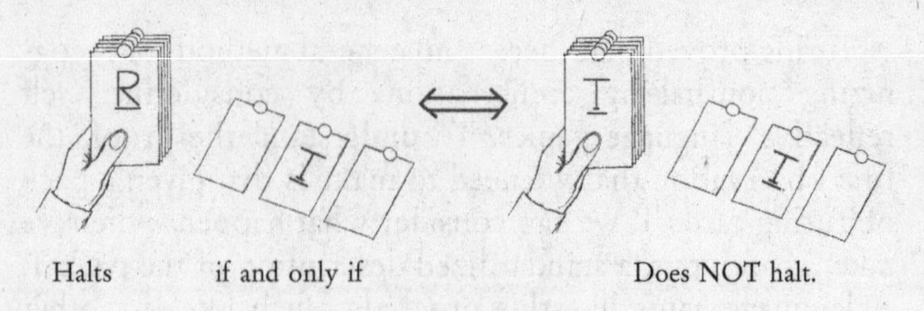

But in that case, what would happen if we used a description of the pack R as our chosen input T?

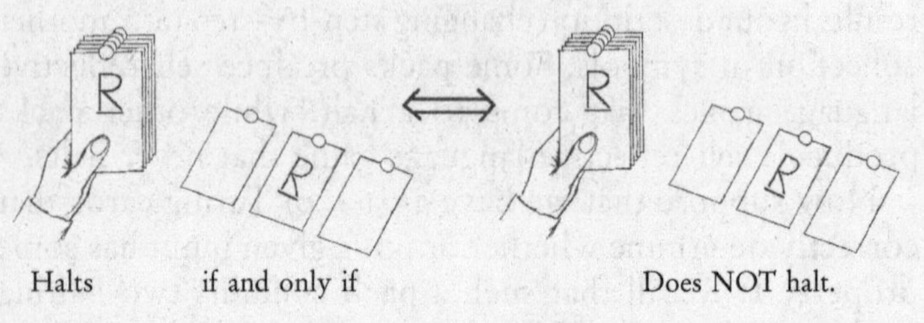

We have two copies of the same game, and if the first version of the game reaches a halting card, the second copy cannot ever reach a halting card. Similarly, if the first copy does NOT halt, that very same game must halt!

Our argument has hit a paradox. The question is, where have we gone wrong? The answer is, we were wrong to assume that a pack like R exists. In other words, the preceding argument compels us to conclude that there cannot be a reliable, automatic method for recognizing when a given Turing pack T produces a non-halting, self-reflecting game. Given that we cannot find a general method for recognizing when a self-reflecting game will halt, we certainly cannot find a general method for identifying when a given input and a given pack of Turing cards results in an unending computation, as any method for identifying non-halting computations could be used to identify non-halting self-reflecting games. In other words, Turing's argument proves that we cannot always recognize when a given computational procedure will never come to a halt.

This is a very important observation, as among other things, it shows that there are problems which cannot be solved using a finite, pre-stated system.

Figure and Ground

We have seen that a decision problem is a kind of challenge, where we consider a collection of things, and we are asked to decide whether or not some arbitrary individual has a particular property. For example, we might consider the set of integers, and ask the decision problem 'Is this integer prime?' If we can correctly answer a decision problem 'yes' or 'no' by using a pack of Turing cards, we say that the problem is decidable. Similarly, we say that a problem is semi-decidable if we can recognize when a thing has property P, but there is no general method for recognizing when a thing does NOT have property P.

The existence of semi-decidable problems surprised many mathematicians, who had implicitly assumed that any definition for a figure necessarily contains the same information as a definition for the ground (i.e. the points which are inside the frame of reference, but not inside the figure).

Some patterns have a constructive definition, and can be generated by a finite system of rules. The process of generating a given pattern may resemble counting, in the sense that it extends to infinity, but every part of the pattern is completed after a finite amount of time. The somewhat surprising fact is that the existence of a constructive definition for a pattern A does not imply the existence of a corresponding, constructive definition for NOT A. Also notice that the word NOT is being used in context with a predetermined sense of thing, as is represented by the box surrounding the previous drawing. For example, A might be the set of square integers, which would make NOT A the set of integers that are not square, rather than the set of anything whatsoever that isn't a square number.

Of course, there is an easy and finite test for finding out whether or not an integer is square. However, things get much more interesting when we consider the general case of adding and multiplying an arbitrary string of integer variables. For example, we can consider the following examples of 'Diophantine' equations, which take an integer or integers as input, and produce an integer output:

$$p(x) = x^3,\ p(x) = x^2 - 4 \text{ or } p(x,y) = 7xy^5 - 3x^2y^3.$$

Diophantine equations are named after the mathematician Diophantus (*c.* 210–294), and they are basically just polynomials, except that in a Diophantine equation, all of the terms must be integers. In the next chapter we will examine one of the deepest questions in mathematics: 'Which Diophantine equations have integer solutions?' To get some sense of why this question has such depth, imagine trying to devise a method for sorting Diophantine equations into those that have an integer solution, and those that do not. If a Diophantine equation has an integer solution, we can certainly recognize that fact. For example, $p(x) = x^2 - 4$ is equal to zero when $x = 2$, and we can be sure that $x = 2$ is indeed a solution. Furthermore, the process of finding such an integer, and proving that it is indeed a solution, is an entirely computable procedure. Theoretically speaking we can find the solution of *any* solvable Diophantine equation just by trial and error: we can simply try every integer in turn, as proving that a putative solution is indeed a solution only requires the ability to add and multiply integers correctly.

Since prehistoric times, mathematicians have explored the solutions to polynomial equations. For example, the Ancient Babylonians knew how to solve quadratic equations, and across the globe, over three or four millennia, many other sophisticated techniques have been developed. At this point, the important thing to note is that if a solvable equation has a solution, the solution itself stands as proof that the equation has a solution. In contrast, consider an equation that does not have a solution. What kind of proof might we provide to show that the given equation has no solution? What kind of statement can stand witness to this kind of truth?

Semi-Decidable Problems

Recall that a property P is said to be 'decidable over a set S' if it is theoretically possible to calculate whether or not any given member of S has the property P. For example, we say that the property of being even is decidable over the integers because a single finite program can be used to determine whether or not an arbitrary integer is even. More interestingly, a property P is said to be semi-decidable over a set S if given any member of S that has the property P, there is a finite calculation that confirms the fact that the given member of S does indeed have the property P. For example, the property of 'having an integer solution' is semi-decidable over the set of all Diophantine equations, because it is always possible to prove that a solvable equation does indeed have a solution.

Some problems are semi-decidable but not decidable. For example, if we know how to use a pack of Turing cards, we can certainly recognize that a given input and a given pack ultimately produce a halting computation. However, as Turing proved, there is no finite program for recognizing when a given calculation will never halt. Remarkably, every problem that is semi-decidable but not decidable has to have certain characteristic features. To appreciate one of the characteristic features of semi-decidable problems, imagine that we have a property P and a machine that says 'yes' if and only if we input a thing with property P. Furthermore, suppose that we can look at any input I and come up with a number *t*(I), such that if the machine is going to answer 'yes', we can be certain that it will do so in less than *t*(I) units of time. In that case we could take any input I, calculate the corresponding number *t*(I), plug our input into the machine and wait for one of two things to happen:

1. The machine recognizes that the input I has the property P.
2. A period of time *t*(I) passes, and our machine still hasn't said anything.

In the second case we know that our input does NOT have property P, simply because the calculation has taken too long. This observation tells us that the property in question must be decidable, because we can recognize the absence of P by the silence of our machine. This means that if a property P is semi-decidable but not decidable, there cannot be a way of finding a safe overestimate for the time required to confirm that a given input has the property P. By definition of semi-decidability we will eventually be able to confirm that a given input has property P, but in general we cannot know how many computational steps will be required to establish this fact.

Another necessary property of semi-decidable sets is the following. Suppose that our set of things can be written out as a particular list. For example, imagine listing the integers in ascending order. If we can move along this list ticking off EVERY thing with property P, then EVERY thing we pass over without ticking must, by definition, NOT have property P.

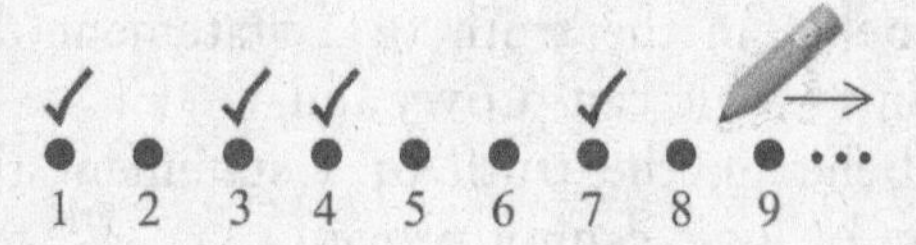

If property P is decidable, we can systematically move along our list ticking the things that have the property P. Conversely, if we can systematically tick off elements in the order they appear in some list, then property P must be decidable. This follows because if we could tick off the things with property P in order, we would have a method for recognizing things that do NOT have property P, namely looking for things that have not been ticked, but which are earlier in the list than something that has been ticked. In other words, if a property P is semi-decidable but not decidable, then for every list and every method for ticking things with property P, our ticking pencil must hop around an infinite amount, moving backwards as well as forward.

Returning to a particular example of a semi-decidable problem, let's consider Turing's halting problem. In that example of a decision problem, we ask: 'If I use these Turing cards on that input, will my computation ever halt?' In terms of logical truth, we are inclined to say that an input/Turing card combination either halts or does NOT halt. Furthermore, if an input/Turing card combination halts we can prove that it halts, simply by carrying out the calculation. On the other hand, just because we cannot *prove* that a given calculation halts, that doesn't necessarily imply the existence of a proof that the given calculation never halts. In short, there are cases where we cannot prove a statement that we believe to be true.

Perceiving this gulf between the formally demonstrable and the statements we accept as truths leads us to consider the following question. What is the minimum difference between a justification for:

1. Our belief in the truth of a statement that our system of logic can prove, and
2. Our belief in the truth of a statement that our system of logic cannot prove.

If we could show that there is always a difference between such justifications, we might come to the conclusion that we should drop the concept of unprovable truth altogether. In other words, could it be the case that only provable truths are really true?

Chapter 11:

KURT GÖDEL AND THE POWER OF POLYNOMIALS

> 'The utility of systems lies not merely with their making us think about something in an ordered way according to a particular scheme, but in making us think about it at all; the latter utility is incontestably greater than the former.'
>
> Georg Christoph Lichtenberg, 1742–1799

Matiyasevich's Theorem

Before we return to the concept of unprovable truth, we need to sketch out a beautiful proof of Gödel's Incompleteness Theorems. More specifically, we shall prove Gödel's famous theorems by means of another result of staggering significance, called Matiyasevich's Theorem (also known as the MPDR Theorem). This less well-known theorem has its roots in the 1960s, and the work of the American mathematicians Julia Robinson, Martin Davis and Hilary Putnam. In effect, they made progress by applying modern mathematical logic to the study of Diophantine equations. More specifically, they were interested in the logical properties of Diophantine definitions. A Diophantine definition is simply a standard way to define a set of integers, by using a polynomial equation.

For example, the even numbers can be defined as the set of integers n for which the following statement is true:

'For SOME integer m, $n-2m=0$.'

Similarly, the square numbers can be defined as the set of integers n where:

'For SOME integer m, $n-m^2=0$.'

As I have already mentioned, if someone claims to have found some integers that solve an equation P, it is very easy to put that claim to the test. All you need is a definition of the equation P, plus a few simple rules that tell you how to add, subtract and multiply. In fact, all we really need are the following five axioms (due to Giuseppe Peano, 1858–1932):

$$n+(-n)=0,\ n+0=n,\ n+(m+1)=(n+m)+1,$$

$$n\times 0=0 \text{ and } n\times(m+1)=(n\times m)+m.$$

Now, suppose that we are considering some particular Diophantine equation P. If there is a list of integers $n, m_1, m_2, \ldots, m_k$ that solve the equation P, we can definitely find that list in a finite amount of time. Even if we don't have a clever method for finding a solution, we could simply try every possible combination of integers in turn. The number of steps in this trial-and-error process might be vastly greater than the number of electrons in the universe, but so long as it is finite, we say that the procedure is computable.

The logical structure of Diophantine definitions is admirably clear, because in a more or less efficient way we can set about looking for solutions of the form $n, m_1, m_2, \ldots, m_k$, and in the case where such a solution exists, we know by definition that the integer n has the property P. The integers $m_1, m_2, \ldots, m_k$ are somewhat analogous to the working out that a Turing pack might instruct us to write: they are an essential part the calculation, but they aren't part of the final answer.

Up until 1970, only a few mathematicians suspected that Diophantine definitions might be every bit as powerful as Turing cards. In other words, only a few people guessed that the addition and multiplication of integers might literally generate the entire computable universe. The proof that Diophantine equations are a kind of universal programming language proceeded in a number of steps. By 1970, recursion theorists could prove that Diophantine sets had almost all of the necessary properties. In particular, Julia Robinson, Martin Davis and Hilary Putnam had established the following facts:

1. Every finite list $a, b, \ldots, z$ has a Diophantine definition. It is of the form $(a-n)(b-n) \ldots (z-n)=0$, so the number n belongs on the list if and only if it is one of the numbers $a, b, \ldots, z$.
2. If the sets A and B have Diophantine definitions, so does A OR B. If $P(n, x)=0$ and $Q(n, y)=0$ are the respective Diophantine definitions for the sets A and B, then $P(n, x) \times Q(n, y)=0$ is a Diophantine definition for A OR B. For example, $(n-2x)(n-y^2)=0$ picks out the integers n, which are either even or square.
3. Similarly, $P(n, x)^2 + Q(n, y)^2 = 0$ is a Diophantine definition for A AND B.

Only one remaining property (called bounded universal quantification) remained elusive. Recursion theorists realized that if they could only establish the existence of a single Diophantine set whose members grew exponentially (like 1, 10, 100, 1 000, ...), then every definable set would necessarily have a Diophantine definition.

Most people thought that no such set could exist, but on 4 January 1970, a twenty-two-year-old Russian mathematician by the name of Yuri Matiyasevich proved the doubters wrong. The complete proof of Matiyasevich's Theorem is much too involved for a book of this size or nature, but roughly speaking his argument can be divided into two parts. First, a parametric version of the Fibonacci sequence is shown to have a wide range of properties, each of which has a Diophantine definition. Second, he showed that the combination of these Diophantine properties suffice to specify completely the exponential sequence in question.

The fact that the prime numbers have a Diophantine definition was a considerable surprise to many mathematicians, but it really is a provable fact that multi-variable polynomials have such a wide range of different solution sets, that every definable set of integers has a corresponding polynomial. Furthermore, as there are no major difficulties in transliterating from one alphabet to another, we can translate any collection of symbols into a sequence of digits. This implies that every definable set of symbols has a Diophantine definition. It is literally true to say that if there is a finite, deterministic rule for generating a sequence of symbols (even an infinite sequence of symbols), then there must be a rule that generates those same symbols simply by adding and multiplying.

A remarkable implication of this result can be found by considering a 'universal decision pack'. A player using such a pack receives descriptions of decision packs as inputs, together with a particular case that might or might not have the property in question. The universal decision pack reads the input pack P together with the particular input N, and works to recognize whether or not the input N has the property that is recognized by the decision pack P. Since it does this in an entirely deterministic manner, we have a well-defined subset of the set of all pack-input pairs, namely those where the input does indeed have the property encoded by the pack.

Given any particular alphabet that we might be using, every possible input of symbols can be assigned a unique, identifying number, because we can list all the possible inputs alphabetically. I shall use the symbol n to denote the integer that identifies a given input. Similarly, every property can be assigned a unique number, because we can list our descriptions of decision packs in alphabetical order. I shall use the symbol p to denote the integer that identifies a given decision pack. Matiyasevich's Theorem implies that for every universal decision pack, there must be an equivalent, Diophantine equation $U(n, p, x)$. In other words, the polynomial '$U(n, p, x) = 0$' has a solution if and only if the input n has the property p.

The polynomial U is just an ordinary polynomial, involving nothing more complicated than the addition and multiplication of some integer-valued constants with two integer-valued parameters (n and p) and some fixed number of integer-valued variables, which I have denoted x. They tend to look rather messy when written out in full, but the crucial point is that equations like U actually exist, and they are called universal equations because *by altering a single, integer-valued parameter p we can use the equation* U *to generate every definable set of integers.* In other words, the solution sets of a universal equation literally contain the entire computable universe!

If we are oblivious to calculation time, it does not take much to attain the greatest level of computational power. Even the basic rules of addition and multiplication can ultimately generate any pattern whatsoever, provided that the pattern can be computed in a deterministic or mechanical manner. Also notice that the patterns generated by a universal pack of Turing cards, or a universal equation, are arbitrarily complex. That is to say, because the system is universal, it can replicate the simplest patterns, the most complicated patterns, and everything in between. A similar observation holds for every other quality that a definable list of numbers can be said to possess.

Kurt Gödel

Kurt Gödel (1906–1978) was born in Vienna, and when he was six years old he contracted rheumatic fever. He recovered to full health, but when he was eight he began reading medical texts on the illness he had suffered, and became convinced that he had a weak heart. This unfounded suspicion was the beginning of a lifelong obsession with his health; an obsession that became particularly problematic when Gödel concluded that food was fraught with danger, so people should eat as little as possible. As a student at the University of Vienna, he was persuaded to move from physics to mathematical logic by the gifted Philip Furtwängler.

Furtwängler had to dictate everything he wrote, because he was paralyzed from the neck down. The young Gödel was greatly impressed by the image of a mind exploring the world of numbers, unhindered by the lack of a working body. A lifelong believer in a supernatural creator, Gödel was also a fervent Platonist. In other words, he believed that abstract, mathematical objects have a very real existence, independent from mathematicians, or the languages that mathematicians can use. He was by all accounts a brilliant and fastidious man, and he seems to have been one of those people who are drawn to maths not only because of the intellectual challenge, but also because of the unworldly, timeless and uncontaminated character of its objects of study.

Gödel was disgusted by the Nazis, but even as war was looming, he did not want to flee his native Vienna. When the Second World War finally broke out, Gödel was judged fit for military service, and he greatly feared being conscripted into the German army. He finally decided to flee Austria, and in 1940 he arrived in America. As a world famous mathematician he was offered a prestigious post at the Institute for Advanced Study in Princeton. Here he became close friends with a man he had first met in 1933: Albert Einstein. In a letter to his mother, Gödel wrote that the two men would meet daily at Einstein's house, leaving at ten or eleven in the morning, and walking for half an hour until they reached the institute. They would then meet up again at one or two, before walking home together, chatting in their native German. It is clear that both men enjoyed their discussions on politics, philosophy and physics, and towards the end of his life Einstein even remarked that he only bothered going to his office 'to have the privilege of walking home with Kurt Gödel'.

Gödel was always drawn to problems of philosophical interest, and in 1949 he made a remarkable foray into the theory of relativity. More specifically, he showed that there are solutions to Einstein's equations that contain closed loops. In other words, he showed that if you have enough energy, time travel becomes a theoretical possibility! Gödel was greatly saddened when his friend Einstein died in 1955, and as he grew older, Gödel became increasingly concerned about his own physical condition. When his beloved wife Adele suffered some serious health problems, his difficulties with food and paranoia over poisoning became even more acute. By the end of his life he was so frightened of being poisoned that he refused to eat at all, and in 1978 he effectively starved himself to death.

Although he made many important contributions to mathematics and logic, Gödel is most famous for proving an intellectual bombshell: the Incompleteness of Arithmetic. A formal system is said to be complete if every grammatically correct statement in the given language can either be proved or disproved from the axioms of the system. For example, if we restrict the language of arithmetic by refusing to use the words 'EVERY' and 'SOME', the fragment that remains is in fact complete. In other words, the few mathematical statements that we are left with can all be proved or disproved in a simple, mechanical manner, using a small number of fundamental axioms, together with some basic rules of deduction. As we shall see, the situation is much richer and more interesting when we include the words 'EVERY' and 'SOME' into the vocabulary of maths.

Searching for Solutions

Imagine a nation of people who prize the study of Diophantine equations above all else. In the middle of their kingdom stands an imposing stone monolith, covered in mathematics. If the people knew that a statement was true of the integers, they might carve it onto this monolith. Over the generations many statements had been added, but the most revered carvings were the laws for the addition and multiplication of integers (e.g. Peano's axioms), together with the laws of logic. Other truths might also be added, but these people wouldn't carve any old statement onto their sacred stone.

When two men wanted to marry the same woman, she would pick a Diophantine equation, and assign one of the men 'solvable', and the other 'NOT solvable'. The suitor who was assigned 'solvable' would immediately set about trying to find a solution, using the sacred instructions for adding and multiplying integers. The suitor who was assigned 'NOT solvable' had a somewhat more interesting task.

It is easy to prove that some equations have no integer solution. For example, $x^2+1=0$ has no integer solution, because for every integer x, x^2 is at least as big as zero. The monolith would not need many axioms to be powerful enough to prove that there is no integer n such that $x^2+1=0$. Other Diophantine equations require more sophisticated methods to prove that they have no solution.

When a suitor thinks that the given equation has no solution, he has to find something on the monolith that tells him that his argument is valid (e.g. the monolith ought to confirm the validity of the preceding argument). If nothing on the monolith confirms the validity of a suitor's argument, the only remaining option is to persuade the King to carve another statement onto the monolith. This was considered a very serious business, because if the King allowed inconsistent statements onto the monolith, the logic sanctioned by the sacred stone would be utterly useless. Among other problems, the people would face the embarrassing situation in which both suitors would invariably win the right to marry.

One day, the King's daughter (an exceptional beauty named Helen) was compelled to pick a Diophantine equation. It took her ten minutes to dictate the monstrous thing, and the poor suitors were quite ashen by the time she assigned them their respective tasks. The young man Kurt was assigned the 'NOT solvable' task, while Bill got 'solvable'. Bill was a bit of a computer whiz, and by the end of the day he had checked that there were no solutions with less than a billion digits. Meanwhile, Kurt grabbed a pencil and tried to look for a pattern ...

The Incompleteness of Arithmetic

Recall that axiomatic systems and the laws of logic can be summarized on a computer program, and hence they can also be summarized in Diophantine form. The fundamental proof theoretic fact is that we can take any set of axioms, or any countable axiom scheme with a finite definition, and devise a program that finds each logical consequence in turn. In particular, we can use such a system to find proofs that might be used by a suitor whose task is to demonstrate that a given equation doesn't have any solutions. Assuming that no extra axioms are carved onto the monolith, this is a perfectly mechanical procedure. Therefore Church's thesis implies that we can summarize this procedure using a pack of Turing cards.

Furthermore, we can list all the Diophantine equations in alphabetical order, which is to say that we can associate each equation with a unique number. Some numbers will correspond to equations that can be mechanically proved to be unsolvable. For example, we might use the monolith to prove that $x^2+1=0$ has no integer solution, $x^2+2=0$ has no integer solution, and so on. Each of these equations has a unique, identifying number, so our mechanical procedure identifies a subset of the integers, namely those integers that correspond to an equation p where we can mechanically prove that p has no solution. Matiyasevich's Theorem implies that this subset of the integers must have a Diophantine definition. With that in mind, we are ready to return to the story of Kurt and the King's daughter.

Kurt soon realized that Helen's equation was related to a universal Diophantine equation, and he spent many months studying it, considering things that the integers might encode. He was convinced that the King's daughter (who was herself a formidable mathematician) must have picked those numbers for a reason. The more he looked, the more he found, and eventually he picked out substructures that encoded everything on the village monolith. The final breakthrough came when he tried converting one of the constants into a variable. He immediately recognized the equation that this produced, because it captured the property of being 'self-reflectively unsolvable'.

An integer n is said to be self-reflectively unsolvable for U if and only if you can use the village monolith to prove that $U(n, n, x)=0$ has no integer solutions. For example, if $U(1, 1, x)=x^2+1$, we would say that 1 is self-reflectively unsolvable for U (assuming that our axiom system can prove that $x^2+1=0$ has no integer solution). Notice that there are two facts involved. The first fact is that when we set the two parameters of our universal equation equal to 1, the resulting equation has no integer solution. The second fact is that we can prove that this equation has no integer solution just by using the axioms on the monolith. The crucial point is that the formal procedure that demonstrates this second fact can be encoded into a pack of Turing cards, or into a Diophantine equation.

Kurt could hardly contain his excitement as he calculated exactly which value for n Helen had used to produce her equation. As he examined this enormous number, a sense of *déjà vu* crept over him. A quick calculation confirmed his suspicions: the number Helen had used encoded a Diophantine equation which exactly mirrored the property he had just been studying!

The point of this story is that Kurt is looking at an equation that necessarily exists. That is to say, for every system of axioms (which correctly adds and multiplies integers), there is some Diophantine equation U and some integer T with the following, remarkable property:

For every integer n,

1. $U(T, n, x) = 0$ has a solution if and only if
2. The monolith is powerful enough to prove that $U(n, n, x) = 0$ has no solutions.

To repeat, we can be certain that the two statements above are logically equivalent, and this is the heart of Gödel's Incompleteness Theorems. We know that it is true because given the axiomatic system in play, there must be some integer T that corresponds to a program for finding proofs of the following kind of statement: '$U(n, n, x) = 0$ has no solution.' Note that n is some integer which we are given, while x is a finite string of integer-valued variables. Helen could have picked any integer n to plug into the equation U, but she decided to use T, the integer which corresponds to the procedure for finding proofs of statements of the form '$U(n, n, x) = 0$ has no solution'.

Kurt took a deep breath, and considered what would

happen if statements 1 and 2 were true for the particular case when $n = T$. Statement 1 would tell us that $U(T, T, x) = 0$ has a solution, so we should be able to prove this 'fact' using our axioms for addition and multiplication. The logically equivalent statement 2 tells us that we can prove that $U(T, T, x) = 0$ does NOT have a solution. In other words, if Helen's equation $U(T, T, x) = 0$ has a solution, then the monolith must be inconsistent.

Returning to statements 1 and 2, if we put $n = T$ there are two conceivable cases: $U(T, T, x) = 0$ either has a solution, or it doesn't. If we suppose that $U(T, T, x) = 0$ has a solution, then our PC logic machine will be able to use the monolith's axioms to generate the following statements:

1. $U(T, T, x) = 0$ has an integer solution
   and
2. The monolith is powerful enough to prove that $U(T, T, x) = 0$ has no integer solution.

In this case we would have conclusive proof that the axioms on the monolith are inconsistent, as we would be able to prove that $U(T, T, x) = 0$ has a solution AND $U(T, T, x) = 0$ does NOT have a solution. But what if $U(T, T, x) = 0$ does not have a solution? In other words, what happens when statements 1 and 2 are false? In that case we manage to avoid paradox by observing the gap between truth and proof:

NOT 1 $U(T, T, x) = 0$ has no solution
and
NOT 2 The monolith is not powerful enough to prove that $U(T, T, x) = 0$ has no solutions.

If the monolith is indeed consistent, then Kurt will have to have something added to it before he can complete his task.

He can be quite certain that in its current state, he cannot use it to prove that Helen's equation has no solution. As things stand, Kurt is quite certain that he cannot complete his allotted task. Indeed, for every consistent set of axioms there are infinitely many true arithmetic statements that we cannot prove using nothing but the given axioms. That is what is meant by the Incompleteness of Arithmetic.

Once he had completed this argument, Kurt headed straight to the palace. As he waited for an audience with the King, he wondered how to handle his peculiar predicament. What axiom would he need to carve before he could complete his task? And how would he persuade the King to add it to the monolith?

Kurt: Tell me, your majesty, do you think that your daughter should marry both her suitors?

King: How dare you suggest such a scandalous thing! I should have you flogged for asking such an impertinent question.

Kurt: I do apologize, I meant no offence. I agree that the monolith is entirely consistent, and such a thing is quite impossible. Unfortunately, I can't find anything on the monolith that tells me that.

King: What kind of madness is this? You don't need the monolith to tell you that its axioms are consistent. If my ancestors so much as suspected that one of the axioms was not consistent with the others, they would never have allowed it to be carved onto the monolith.

Kurt: Does that mean I am allowed to make that assumption, in my proof of unsolvability?

King: I don't see why not. Every proof implicitly assumes that the monolith is consistent – that assumption is the very basis for our belief in the truth of the things we prove. After all, if our axioms are inconsistent, then our theories are worth no more than any other sentence! It is decided then – I hereby decree, by royal command, that you may assume that one and only one person has the right to marry my daughter.

Kurt: In that case you had better fetch her. If she has to marry one and only one person, then that person must be me.

Truth, Proof and Consistency

People have always argued about the nature of truth, but in some ways everybody understands the meaning of the word. People say that a statement is 'true' if and only if our understanding of the statement accords with our understanding of the subject matter or situation that the statement is about. Of course, we sometimes think that a statement is true, and then realize it is not. We therefore understand that a statement is only true if it can withstand every kind of relevant scrutiny, and we generally don't know all the different forms of scrutiny that could possibly show that our statement is false. That is why it is hard to know when a statement is really true, and why a 'true' statement is more than a just a well justified statement. I also think it is worth emphasizing that reality is present whether we can talk about it or not, but 'truth' is fundamentally a relationship between language and the world. Crucially, it seems reasonable to say that when we subject potentially true statements to every relevant form of scrutiny, we need only consider the kinds of scrutiny that might be brought to bear by people (real or idealized) who understand the language that we use when we make our statements.

In the case of mathematics our choice of axioms largely

determines the meaning of our symbols. For example, if I make a provable statement using the language of axiomatic set theory, it doesn't make sense to say that my statement is actually false simply because you refuse to accept one of my axioms. The axioms are part of what makes my symbols mean what they mean, and if you don't accept my statement simply because you don't accept one of my axioms, that does not show that my statement (as originally conceived) is actually false. It simply shows that you understand my symbols differently from me, and at best your argument might persuade me that it is a good idea to adopt some new axioms. In that sense there really are mathematical statements that are true by definition. Likewise, we are the ones who determine the rules of chess, and we really can know for certain that you cannot force checkmate against a lone king with only a king and a pair of knights. This kind of truth by definition is what makes it possible to construct proofs in the first place, but that is not the only kind of mathematical truth!

For example, I do not think that '$1+1=2$' is merely true by definition. It is also true in the usual, less certain sense, as *we can understand that this statement fits with the way that we understand the relevant subject matters*, which are the concepts of number and addition. After all, if you have a single object in your collection and you add another object, you really do have two objects in your collection. Likewise, if you take one step forward and then take another step forward, you really have taken two steps forward. These deeply understandable experiences are part and parcel of the concepts of number and addition, and you don't understand my language if you don't understand that fact. We can see that the statement '$1+1=2$' fits with our common understanding of the concepts of number and addition, and in that sense we can see that it is true. Similarly, the statement '$n+0=n$' is true in the sense that if I take n steps forward and then don't take any more steps, I have in fact taken n steps forward. It is also true by definition, as this statement is one of the axioms governing the proper use of the symbol '0'.

Of course, most mathematical statements are far too complex for us to assess in such an intuitive manner. Nevertheless, it is this kind of truth, or fidelity to an underlying concept, that leads us to accept certain axioms in the first place, making proof and 'truth by definition' possible. Mathematicians need to employ rule-governed systems in order to do their work, but the question then arises as to why we should trust those systems once they start to outstrip our intuitions. In particular, how can we be sure that our rule-governed, symbolic systems really are consistent, and cannot lead us to contradict something that we ought to accept as true? It is these mysteries that Gödel helped to elucidate, and his insights came as a surprise.

The first point to make is that some mathematical systems really can prove their own consistency. For example, the axioms of classical logic can be used to prove the consistency of classical logic. That is to say, there is no way that we can start with a statement A, apply the laws of logic, and reach the conclusion NOT A. Some people argue that there are better systems of logic than the classical one, but there is no question that the language of AND, OR and NOT is logically consistent.

On the other hand, arithmetic cannot prove its own consistency. We can prove that fact because we know that for every formal system capable of adding and multiplying integers, there is an equation that corresponds to $U(T, T, x) = 0$. In other words, for every axiom system there is an equation that has a solution if and only if we can prove that the equation has no solution. This means that if our axiom system is consistent, the equation cannot have a solution. If our formal system could prove its own consistency, we could also prove that the equation U(T,

T, x) = 0 has no solutions. But if we can prove that it does not have solution, we know that it must have a solution, which is a fact that we can prove! In other words, arithmetic axioms that can prove their own consistency must be inconsistent.

In light of this fact, we must be exceptionally careful with formal assertions that are equivalent to the King's decree. We can claim that our formal system is consistent, but we cannot make that claim an explicit feature of the formal system itself. However, there is at least one way in which we can consistently increase the power of the monolith in light of the insight provided by Kurt's formal argument. At the very least, we can carve the following statement onto the monolith: 'U(T, T, x) = 0 does NOT have a solution.'

Adding this statement enables us to prove at least one new thing (namely the statement itself). Furthermore, if our new expanded system is inconsistent, then so was our old system, because if you can prove that 'U(T, T, x) = 0 has a solution', you not only contradict our new axiom, you also demonstrate an inconsistency in the original axiom system. The crucial point is that we can extend our axiom system by adding the statement 'U(T, T, x) = 0 does NOT have a solution', and we might have a good reason for wanting to do that. In contrast, we cannot consistently extend our axiom system by adding the statement 'U(T, T, x) = 0 has a solution'. If we accept that our original system was consistent, then as members of the logical community we must also accept that 'U(T, T, x) = 0 does NOT have a solution', even though we cannot prove that statement without adding a new axiom.

Once we have increased the monolith's power in this fashion, the set of provably unsolvable Diophantine equations becomes slightly larger. The power of an axiom system can always be increased further in this manner, but this procedure is not itself computable. Each step requires a further insight, as we must reconstruct Gödel's argument in the context of a new set of axioms. Also note that we can have good reasons for adding such a statement to our axioms even though this statement is not a self-evident truth. The evidence or justification for this truth is found in our comprehension of the other axioms, together with Gödel's argument (the context of our statement).

Our statement is logically independent from its context (it is saying something new), but in some sense it is a natural extension of the prior system. We can see that 'U(T, T, x)=0 does NOT have a solution' is true by following the previous arguments, and by understanding the axioms on the monolith. On the other hand, a mechanical application of those same axioms cannot recognize this truth. That is to say, if Kurt had used a PC machine instead of his intellect, he could not have completed his task. This is what I meant when I said that mathematics is not a deterministic language game. However, it is not clear that this way of extending our axioms requires an inspiration that is beyond any finite description of a formal methodology. We can't use a single program to complete our set of axioms, but we may be able to give strategic advice that covers the relevant cases. By way of analogy, a person may know no numbers but the integers, and using the language of integers we can ask such a person 'if $2x=1$, what is x equal to?' To answer this question we need an expanded language that incorporates fractions, but this question can show the sense in extending our language, and persuade our audience that fractions make sense.

We could summarize Gödel's achievement by saying that he had the brilliant idea of using a very formal, mechanical proof to show the limits of formal, mechanical proofs. Make no mistake: he did not discover some mysterious truth that formal language can never articulate. Similarly, he did not prove that there are arguments that a human being can follow but which no computer could ever construct. What he showed is that for any particular formal system (such as a computer might use) there is a Diophantine equation with no integer solution, though the given computer program cannot prove that this equation has no integer solution. Another computer might be perfectly capable of proving that the equation has no solution, but in that case, there must be a second unsolvable Diophantine equation, which falls through the web of proof for our second computer. No matter how many formal principles we cobble together, there will always be some Diophantine equation whose unsolvability we cannot prove.

In an earlier section I described the incompleteness of arithmetic as an intellectual bombshell. The grand project that it really blew apart was known as 'Hilbert's program': the dream that all of mathematics could be formalized within a finite, provably consistent axiomatic system. Perhaps the most extreme example of logically explicit mathematics was Russell and Whitehead's *Principia Mathematica* (published in 1910, 1912 and 1913), which dotted every 'i' and crossed every 't' for the bulk of ordinary mathematics. Indeed, the authors were so careful to spell out every logical assumption that it took them 362 pages to build up enough machinery to reach the conclusion that '1 + 1 = 2'!

The logical system employed by Russell and Whitehead was painstakingly spelled out, and in a sense their book was a precursor to modern programming languages. Everyone, including Gödel, was confident that all of their conclusions were correct. The problem was that Gödel proved that the logical system specified at the beginning of the book couldn't possibly prove every theorem that the authors hoped it might prove. What is worse, he showed that no amount of tinkering with the axioms would help: there will never be a complete, definitive list of everything you need to know in order to prove every arithmetic statement that we think is true.

At this point, we need to reconsider the statement 'Only provable truths are really true'. We believe in provable truths because we have good reason to trust our axioms and our systems of argument. In particular, if we recognize that a statement is true because of its proof, there is a foundational belief in the consistency of our axioms, as inconsistent axioms are profoundly useless. However, if we formally state our implicit belief in the consistency of our axioms, we produce an unprovable statement. In this sense, if it isn't 'really true' that our axioms are consistent (because we can't prove it), then our provable truths are not 'really true' either!

Chapter 12: MODELLING THE WORLD

> 'Models in [the life sciences] are not meant to be descriptions, pathetic descriptions, of Nature; they are designed to be accurate descriptions of our pathetic thinking about Nature. ... They are meant to expose assumptions, define expectations and help us devise new tests.'
>
> James Black, 1924–2010

Science and the Uses of Models

Science is a complicated activity, involving the collection of data and the critical evaluation of descriptions of events. It involves a disparate patchwork of methodologies, not some single 'scientific method', but all scientific claims are made to be tested, and we are most interested in general explanations, not mere descriptions of specific cases. Scientific laws are an essential part of science (and they are particularly important in physics), but you can certainly work as a scientist without looking for new laws. Many people over the centuries have been astounded by the fact that scientific laws are mathematical in form, as what does 'pure thought' have to do with messy, contingent reality? The extraordinary success of the scientific enterprise is indeed remarkable, but the fact that science tends to become mathematical is not so surprising if we think that mathematics is the language of patterns, while the job of a scientist is to try and identify regularities or patterns.

We shall return to the question of why science tends to become mathematical, as that is central to understanding the development of mathematical thought. First I want to emphasize that we cannot understand the meaning or significance of a general law without some sense of what that law implies, and working out the implications of a physical law is far from straightforward! For example, it is one thing to know that gravity follows an inverse square law, but understanding the implications of that law is another thing altogether. Indeed, several contemporaries of Newton independently imagined that gravity might follow an inverse square law, but Newton is justly given the credit because only he could deduce how an object would move if it was subject to just such a system of forces. That is to say, we can use Newton's mathematical and conceptual scheme to deduce that planets will follow an elliptic orbit, a projectile will follow a parabolic path, and so forth.

Everyone knows that physical laws are important, but scientists and engineers don't make predictions by simply looking up the relevant rules. The application of physical laws is not a neat, axiomatic process, but an art by which practitioners choose and adapt the techniques that are relevant to modelling a given situation. For example, it is not at all obvious what the Law of Gravity tells us about the motion of a paper dart, so we wouldn't use the motion of such a dart to check if Newton's theory is correct. Evaluating the contexts in which a given formalism is useful or relevant is an essential part of the art of the various sciences, and that discipline is not itself a strictly mathematical one.

Biologists, medical researchers, physicists, engineers, economists, social scientists and many other people make progress in their chosen field by constructing mathematical models. A mathematical model is a logical machine for converting assumptions into conclusions, and it is a striking fact that seemingly esoteric forms of mathematics have enabled countless insights into important, empirical events. Since the objects of mathematics are governed by their own rules, we can scrutinize mathematical claims by their own, internal logic, and can be confident that our conclusions really do follow from the given assumptions.

When we judge a mathematical model we don't just assess the validity of the mathematics. We also assess whether mathematical accounts of actual events are apt descriptions of the events in question, and we do this by subjecting our claims to empirical tests, and by opening our work to public scrutiny. Science is corrigible, and even if we are reasonably confident that we know how something works, we cannot be certain that we have not missed some critical fact. Nevertheless, there is no denying that scientists really can show each other something, despite the fact that our knowledge is always incomplete.

Many qualities can make a model useful or important, but the best models are like the best theories: they show that seemingly unrelated observations can all be explained by a single mechanism. Mathematical models are simplifications, not the complete truth, but in many cases we can use them to obtain answers about what will happen in the physical world. This form of excellence in a model can be measured by a machine, as in the case where we predict how much a beam will bend, and can empirically demonstrate that the beam actually bends by the amount predicted by our model.

We may be awestruck by the power of science, but I do not think that we should be surprised that science is mathematical. In a sense we have nowhere else to look, as scientists require logical and mathematical terminology to make precise, quantitative deductions from the theories that they state. If the mathematics has already been worked out, then finding the prediction of your theory will be a form of calculation. Otherwise, the scientist will need to work as a mathematician, developing mathematical or formal systems that are inspired by the object of study.

One of the most successful and influential models is where we assume that gravity is the only relevant force, and represent a projectile as a mass that occupies a single point in space. We can predict the motion of a cannon ball or a star and a planet, but mathematical models are always simplifications or idealizations, and it isn't always clear when we have missed some pertinent fact. Nevertheless, simple, understandable models that get the basics right are an invaluable source of insight, while more complex models may be just as baffling as the real-world system that we wish to study. The eminent physiologist Denis Noble put it well in *The Rise of Computational Biology*: 'Models are partial representations. Their aim is explanation: to show which features of a system are necessary and sufficient to understand it. So, although we could try to understand cardiac rhythm as the interaction of the few thousand protein types that are in any cell, we can in fact understand most of what we wish to know about pacemaker activity from the interaction of around a dozen protein types. The power of a model lies in identifying what is essential, whereas a complete representation would leave us just as wise, or as ignorant, as before.'

To make a prediction we need to construct a model. Moreover, models shape our understanding of the natural world. As the philosopher Nancy Cartwright convincingly argued in *How the Laws of Physics Lie*, models play a role in physics like that of fables in the moral domain: transforming abstract principles into concrete examples. Essentially, they enable us to present a comprehensible and exemplary case of the relevant rule-governed behaviour. For example, we can use Newton's laws to construct a model of a projectile, and the fact that this model projectile behaves like a real one is a principle reason for accepting Newton's laws. What is more, through their successes, failures or limitations, models can help us to refine or revise our theoretical framework. The way that a model fails capture a given phenomenon can show us our limitations, and suggest a change to the theory that we use. Models can also be used to help design experiments, shaping the kinds of observations or measurements that scientists think of making. In short, a good model does more than fit the data: it helps to clarify the way to think about the object we are modelling.

Order and Chaos

Mathematical systems are intrinsically ordered, and any pattern with a regular structure can be described in the language of maths. Because of this fundamental association between mathematics and order, the existence of a branch of mathematics called 'chaos theory' may appear to be a contradiction in terms. However, when mathematicians study 'chaos', they are studying a system of explicitly stated rules, and not the absence of rules that the word chaos usually denotes. The distinctive and remarkable feature of mathematical chaos is that the rules in question generate behaviours that are inherently difficult to predict. To be more precise, a dynamical system is said to be chaotic if objects obeying the given set of rules always remain within a finite region of space, but you cannot predict exactly where the objects will be in the future, because making a small alteration to an object's location results in a series of movements that are very different to what would have happened if you had not made that change. In other words, chaotic systems are governed by highly sensitive functions, which produce very different outputs on the basis of very similar inputs.

The process of kneading dough is an excellent example

of a chaotic system, as it shows how a very simple rule can be extremely sensitive to the initial condition of the system. As the following diagram illustrates, neighbouring points rapidly become separated by the kneading process, making it inherently difficult to predict where a given speck will end up.

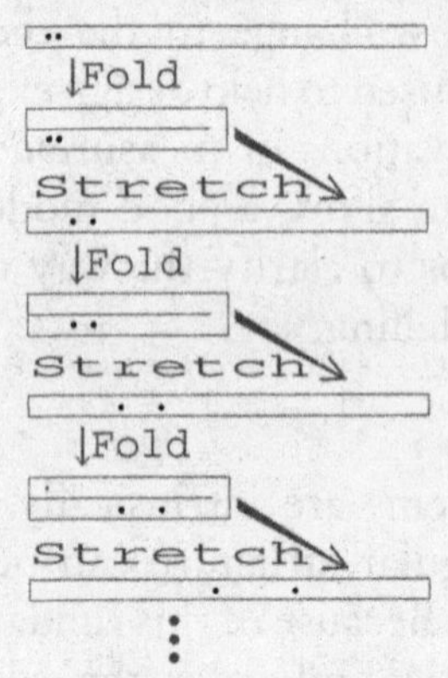

Every time we stretch the dough, the gap between nearby points doubles in length. Any tiny error in our description of a point's original location will grow and grow at an exponential rate. The gap between our estimate and the actual location can double with every kneading motion, which means you need to be arbitrarily accurate to predict the future position of any given point. The fact that we cannot predict the future location of a speck in some kneaded dough demonstrates an important, general principle. Chaos frequently emerges from extremely simple, deterministic programs: a spontaneous emergence of the unpredictable. This fact of life is something of a mixed blessing. On the one hand, it means that even when we know the laws that govern a physical system, we may not be able to predict what will happen. On the other hand, it shows that complex, unpredictable things can be governed by a deep simplicity.

The flow of fluids is a very familiar example of a physical system that can exhibit chaotic behaviour. We find steady and regular movement when fluid is moving slowly, but fast-flowing fluids are turbulent and chaotic. Many systems exhibit more than one kind of behaviour, giving stable or static behaviour for one range of parameter values, periodic or repetitive behaviour for a second set of parameter values, and unpredictable or chaotic behaviour for some third range of parameter values. A beautiful example of a transition from one qualitative type of behaviour to another can be found by blowing over a strip of paper. If we blow gently, the Bernoulli effect lifts the paper to a stable equilibrium, where the lift is balanced with gravity. If we blow a little harder there is a certain point where the paper will overshoot, and as long as we continue to blow at the same rate, it flutters with a regular periodic motion. The boundaries between different, qualitative types of behaviour are called bifurcation points. An example of a bifurcation point is the smallest rate of blowing that will make the paper flutter, as this marks the transition from one qualitative kind of behaviour to another.

A second kind of bifurcation point is where we go from a regular periodic behaviour to a chaotic one, as in the case of a dripping tap. At a low rate of flow a tap will produce a regular drip-drip-drip. If you loosen the tap a little further, you can produce a two-step drip-drop-drip-drop pattern. Further loosening can produce a four-step pattern, followed by an eight-step pattern, and so on. Very rapidly the sequence of drips becomes enormously complex, irregular and inherently difficult to predict, and at a certain point it ceases to be periodic. In practice this is hard to see, because after you reach the chaotic regime, it only takes a little more water for the drops to become a stream. In theory at least, a dripping tap with a low flow rate is ordered and predictable, but as we increase the flow of water the rhythm of the drops becomes chaotic.

That point of transition, from periodic to chaotic, is a second kind of bifurcation point.

Theoretical Biology

Everyone knows that theoretical physicists use a lot of maths. The exceptionally gifted physicist Freeman Dyson put it well in his classic essay 'Mathematics in the Physical Sciences': 'For a physicist mathematics is not just a tool by means of which phenomena can be calculated, it is the main source of concepts and principles by means of which new theories can be created.' In contrast biology is far less mathematical, and the role of mathematics in biological science is much less widely known. Nevertheless, mathematics and computing are increasingly essential to biologists. A great deal of money and effort is being fed into a process whereby mathematical models and experimental data are used to develop ideas and hypotheses, which can then be tested experimentally, leading to the refinement and expansion of the original models.

The first point to make about the relationship between maths and biology is that careful and methodical observation is a central part of every science, and many poor mathematicians have made hugely important contributions to the life sciences. You don't need to be a mathematician to observe the forms and behaviours of living things, but that does not change the fact that mathematics is absolutely fundamental to modern biology. For example, consider the most important idea in biology: Darwin's theory of evolution.

A modern definition of evolution is 'the change in the frequency of the different gene-types (alleles) in a population over time.' The basic idea is that given some population of individuals, the number of individuals may change over time, and the total number and kind of different genes that are present in the population's genomes will also change over time. Some genes will become more or less prevalent, some children will have a slightly different version of a gene compared with their parents (a change from one allele to another), and so on. The word 'evolution' refers to the fact that over time, there are changes in the proportion of the population that carries each of the different alleles.

By this definition, no one can deny that evolution is a real phenomenon. It is like the observation that a dropped object falls towards the ground, not some general theory of gravity that may turn out to be false. Religious fundamentalists may refuse to accept scientific versions of the history of life on earth, but even the most intransigent must surely agree that the frequency of different genes in today's population is different from the frequency of genes in yesteryear's population, if only because the human population has increased from the original two! After all, in a population of two there are only three possible frequencies for any given gene: 0% (neither Adam nor Eve has the gene), 50% (one person has the gene), or 100% (both people have the gene). In contrast there are all manner of different gene frequencies in the population of humans who are alive today, which means that by definition evolution has occurred. More importantly, many of the observed changes in the relative frequency of different alleles can be explained by Darwin's big idea: the process of natural selection. That is to say, if the population in question contains some genetic variation, changes in the frequency of different alleles will occur whenever some genetic types are more likely to survive and reproduce than others.

The concept of frequency or proportion is absolutely essential, as what changes over time is the proportion of individuals with some or other type of genetic inheritance. The point I am trying to make is that we literally could not phrase a theory of evolution without the ability to count. More generally, modern biomedical science involves all manner of sophistical mathematical techniques. There is no question that scientists will continue to use computers and mathematical analysis as a route to understanding biological phenomena, from mathematical descriptions of the interactions between biological molecules, up to the scale of organ physiology, the study of development or the modelling of population dynamics.

The story of how an egg becomes an organism is of particular interest, and to understand such a process we need to consider the genome, and much more besides. In short, organisms are physical entities, and the interactions that are essential to development occur in space and time, subject to the laws of physics. As a simple example, when a mammal produces milk we find spherical droplets of oil suspended in an aqueous fluid. We don't need a gene that says 'droplets in milk should be spherical'. We understand that droplets will be spheres because spheres have the smallest surface area for the given volume, and there are well understood physical reasons why the surface area will minimize.

A moral of this story is that to gain a deeper understanding of the specific details that a biologist might study, it is wise to turn to a more general, mathematical picture of what is going on. As the theoretical biologist Hans Meinhardt argued in *Models of Biological Pattern Formation*, we require theory to make sense from observations. Suppose, for example, that we possessed a perfect biochemical scanner that could tell us the precise concentration of every kind of molecule in every portion of space and time. We might be able to measure the changing concentrations associated with every developmental event, but we would still lack insight as to what is essential to the process, and what was merely incidental. Cause and effect would be jumbled in the data, so although our perfect scanner could tell us a great deal, we would still lack a sense of how the system works. For example, we might not be able to predict what would happen if we managed to change some or other of the measured concentrations.

In order to feel like we understand what is going on, we need to call into being a hypothetical mechanism, or model, which accounts for the data as well as we can manage. Biology is incredibly complicated, and we often make do with an informal word model: an unquantified description of the forces at play. As is the case with any verbal account, the implications and theoretical commitments of such a description are often far from clear. Indeed, at this stage in the development of biomedical science, much of the progress we are making is of the form 'gene X is associated with process Y', or 'brain region X is associated with process Y'. At the risk of sounding unduly critical, observing a correlation between one thing and another barely qualifies as a description of the process in question, let alone an explanation of what is going on. That is not to deny that observing a correlation can be very useful: in some cases information of this kind can even lead to the development of new and effective medical treatments, and it is certainly worth knowing that a given treatment correlates with recovery! It is just that spotting correlations does not provide insight into how the system actually works, unless those observations are helpful in developing or critically evaluating some kind of model, or hypothesized mechanism.

Given our current state of ignorance and the extreme difficulty of measuring all that we might want to measure, it is inevitable that any mathematical model of a biological system is bound to be somewhat arbitrary. Nevertheless, I would claim that precise, mathematical models that can support detailed deductions are superior to informal word models for several reasons:

1. Constructing a mathematical model forces us to be clear about our working assumptions.
2. Proposed biological mechanisms should be consistent with the known principles of physics and chemistry,

and our understanding of those subjects is profoundly mathematical.

3. Systems of interacting parts can behave in counter-intuitive ways, and we need mathematics to be able to identify correctly the implications of our assumptions. In other words, unlike vague descriptions, mathematical models generate clear, unambiguous predictions of both a quantitative and qualitative nature.

Biology is advancing at an awesome pace, but it seems implausible that mathematics will ever dominate biological theory to the extent that it dominates theoretical physics. Nevertheless, many of the recent advances in biology (including the reading of genomes) have required increasingly sophisticated collaborations between experimentalists and mathematically sophisticated theorists. This is a recent development, which partly reflects the fact that unlike their predecessors, modern biologists often deal with very large, computer-based data sets. For example, we cannot hope to make sense of the vast catalogue of genetic data without some sophisticated statistical techniques. However, in addition to providing computational tools and techniques for biomedical research, I hope to persuade the reader that simple mathematic arguments can also play an essential *conceptual* role, as we try to speak aloud the logic of life.

Interactions and Dynamical Systems

Biologists have made huge progress by taking a reductionist approach. For example, we can identify a gene, we can identify the protein that cells produce when they transcribe and translate that gene, and we can look to see how the protein in question behaves. This kind of approach has produced some fantastic science, but there are many biological phenomena that cannot be understood until we put Humpty Dumpty back together again. That is to say, if we are going to understand how a cell or an organism functions, we need to know how the molecular components interact over time.

Mathematical models of complex, dynamical systems are an increasingly central feature of the scientific enterprise. The basic idea is to consider the behaviour of a system of parts, each of which affects its neighbours in accordance with predefined rules. Any particular model of a complex system is likely to have many arbitrary features, but despite the fact that we could have made many other computer models of the same physical phenomena, perspicuously constructed models can explain a great deal. For example, models can explain why buses come in threes, or at least, why buses tend to cluster.

In order to see why buses tend to bunch together, we need to make three simple, empirical observations. The first observation is that it takes time for people to move from a bus stop onto a bus, and since they move onto the bus one at a time, the more people there are at a bus stop, the longer it takes for the bus to pick them all up. The second observation is that the number of people waiting at each bus stop increases over time, as more and more people arrive at the bus stop. Third, people leave the bus stop when a bus arrives. This combination of facts implies that the presence of a second bus some short distance in front of you speeds you up, because the bus in front of you removes people from the stops, so your bus passes the stops more quickly. Similarly, the absence of a bus in front of you slows you down, as the bus stops will contain more people, and it will take longer to pick them all up.

It therefore follows that regular processions of buses are inherently unstable. Small gaps will tend to get smaller as the bus at the back will tend to catch up with the bus in front of it, while conversely, large gaps will tend to get larger. Clusters will naturally emerge, as small disruptions to the regularity of bus arrivals become magnified over time. Bus conductors minimize the clustering effect, as do pre-purchased tickets.

As is the case with all of the models that I sketch out in this chapter, we could be much more specific and quantitative about the assumptions that give form to our model. For example, by being very specific we could use our model to quantify the scale of the clustering effect. However, we do not need to include every detail to understand the principles involved, as questions about the real world can be addressed at different levels of abstraction. This example also demonstrates that not every truth about the world is a mere consequence of 'fundamental' physical laws, as there is more to physics than the rules that govern the behaviour of particles. To put it another way, we can understand why buses tend to cluster, and the fact that the objects in question are composed of atoms is no more relevant than the fact that they are painted red.

Another remarkable model of pattern formation has its roots in one of Alan Turing's best ideas. Back in 1952, Turing was trying to understand the development of an embryo. He knew that when an embryo consists of only two cells, separating those cells results in the growth of identical twins, while leaving them together produces a single individual. Turing was fascinated by the developmental process, and asked himself some difficult questions: How do the two cells 'know' about each other, and how can a group of cells self-organize to create a pattern?

Turing assumed that each cell must 'know' about the other because certain molecules move between them. He imagined an initially uniform chemical mixture, and tried to think how it might spontaneously develop into a pattern with different parts. Being a genius of the first order, he realized that concentration gradients could spontaneously emerge if two substances with different diffusion rates react with one another. This is somewhat counter-intuitive, as diffusion usually smoothes out any differences in concentration. Nevertheless, Turing showed that if the diffusing chemicals are involved in reactions of a given kind, the reaction-diffusion process will create local regions with unusually high (or low) concentrations. Few biologists paid much attention at the time, but the mathematician Alfred Gierer and the theoretical biologist Hans Meinhardt have developed Turing's idea, identifying a key principle in biological pattern formation.

The basic idea is that patterns can be generated by a combination of 'local activation' and 'global inhibition'. There are many different variants of this theme, but as a simple example imagine a flat desert with a few rocks scattered in the sand. If there is no wind, our flat desert could persist over time, and we would not have a pattern of high points and low points. If there is wind, it will blow the sand around. The rocks create a small area that is sheltered from the wind, and sand will accumulate at those points, because it is easier for the sand to be blown into the shelter than it is to be blown out of the shelter. In other words, the presence of wind energy can turn a small rock into a large sand dune. This process of a large thing getting larger, or a high concentration becoming even higher, is what we call local activation.

If local activation was the only effect, we wouldn't get a spatial pattern: we would simply have a concentration (or quantity of sand) that starts small but gets larger and larger and larger. My point is that deserts have high points and low points, and this relates to the fact that there is a finite amount of sand blowing in the wind. If the sand is accumulating in the shelter of the dunes, which means it is not accumulating elsewhere. This long-range reduction in the 'chemical' of interest is what we call global inhibition. As another example of pattern formation by local activation and global inhibition, recall my account of the formation of leaves. Leaves begin to form in the regions of the growing shoot that contain a high concentration of auxin, but in a given stem the concentration of auxin can initially be quite uniform. As cells tend to pump their auxin into the neighbouring cell with the highest auxin concentration, high concentration areas tend to become even higher (local activation), while the rest of the stem is effectively drained of auxin (global inhibition).

Holism and Emergent Phenomena

Organisms, economies and ecosystems all contain large numbers of different, interacting parts. If we want to understand a complex system, having an accurate list of the parts and their properties is certainly important. Nevertheless, it is sometimes the case that knowing lots of details about the individual, isolated parts does not tell you what you want to know about the behaviour of the system. The heart of the matter is that in countless cases, we cannot understand the behaviour of the entire system unless we pay careful attention to how the behaviour of one part influences the behaviour of another.

There are many ways to represent one event being coupled with another, and it is possible to discern some important, general schemes. For example, we are all familiar with the term 'vicious circle' (or its benevolent twin, a 'virtuous circle'), and we have a good idea of when it is appropriate to use this phrase. The notion of a vicious or virtuous circle is closely related to that of 'positive feedback'. Imagine a situation where an increase in the concentration of A triggers an event that produces even more of A, resulting in ever increasing concentrations. The converse to positive feedback is negative feedback, where an increase in A triggers an event that reduces the prevalence of A, while a decrease in A triggers an event that tends to increase the concentration of A. Because low values tend to rise and high values tend to drop, negative feedback can be used to maintain a stable concentration.

It is vital that our bodies' internal environment remains stable, so it is unsurprising that there are many physiological examples of negative feedback. As we all know, if our body temperature rises we cool ourselves down by sweating, and if our body temperature drops we warm ourselves up by shivering. Negative feedback can also be used to produce oscillations, particularly if there are time delays in the system. For example, if I can turn up or turn down the heat in my house, but I am slow to make any adjustments, my house might get very hot before I finally turn down the thermostat. Having turned down the thermostat my house might then become very cold, before I finally get around to turning up the heat again. As a result of these dynamics the temperature in my house might oscillate between hot and cold, and a similar mechanism of negative feedback with time delays can be used to produce oscillations in the chemical concentrations in cells.

There are also important physiological examples of positive feedback. For example, the pituitary gland is located at the base of the brain, and in women it sometimes secretes a small amount of luteinizing hormone (LH). This stimulates the ovaries to secrete oestrogens, and in certain conditions a rise in oestrogen levels in the blood stimulates the pituitary gland to produce more LH, which leads to higher oestrogen levels, which leads to higher LH levels, and so on. Because of this positive feedback a small initial quantity of LH soon results in the production of large concentrations of LH. This phenomenon is known as an LH surge, and it triggers ovulation. Women's blood doesn't always have a high concentration of LH because ovulation temporarily inhibits the ovaries' ability to secrete oestrogens. The resulting drop in oestrogen levels removes the stimulus that produced the initial rise in LH levels, so ovulation allows LH levels to drop back down to the concentrations found before the LH surge.

More and more mathematicians are working with

biologists and other scientists to analyze the behaviour of systems that are comprised of many interacting parts. One of the really fascinating things about studying complex systems is that when we build a model with many interacting parts, we frequently find that it exhibits 'emergent phenomena': forms of behaviour that we cannot possibly predict by looking at any of the individual parts. For example, we cannot understand why buses tend to cluster if we only study an individual bus, and so we might say that the clustering of buses is an emergent phenomenon. By the standards of mathematical science the notion of emergence is somewhat vague, although a wide range of theorists have worked to refine the concept. The practical point is that in many kinds of enquiry, the property or behaviour that we are interested in cannot be made manifest in any single part: to see the thing in question, we need to consider an entire system, including rules for how the parts interact.

Such inherently holistic phenomena can be contrasted with concepts like volume, as it is reasonable to say that a shape's volume precisely consists of the volume of its parts. Interestingly, this claim about the relationship between the whole and the parts of geometric forms cuts to the heart of what is new and different about geometries phrased in the language of sets. For millennia, people agreed with Aristotle when he claimed that the parts of a continuous thing are themselves continuous, so the parts are the same kind of thing as the whole. For example, Aristotle would say that we can identify a point that is inside a sphere, but a sphere is not made of points. As far as he was concerned, you can chop a sphere into smaller and smaller parts, but anything that deserves to be called a part of a sphere must have a finite, non-zero volume.

In contrast, modern mathematicians typically begin by defining a sphere as the set of all points within a given distance from the centre of the sphere. Hence the sphere is conceived as being an infinite set of points, which suggests that the parts of a sphere are points, not volumetric regions. Defining shapes and spaces in terms of sets and points, where each point is identified by a set of real number coordinates, is a very powerful move. It guarantees that we can use algebraic techniques to tackle problems in geometry, but it also leads to counter-intuitive results. In particular, there is a theorem called the Banach-Tarski paradox, which shows that if we chop a sphere into an infinitely fragmented or non-continuous collection of points, then those fragments can be rearranged to make a sphere with a different volume! In other words, all of the points in a small sphere can be rearranged to coincide with all of the points in a large sphere. This is possible because every sphere has the same, infinite number of points, and points themselves have no volume.

Of course, if we chop a sphere into any finite collection of continuous pieces, the total volume of those pieces cannot change: it must always equal the volume of the sphere. In other words, the volume of the whole is equal to the total volume of the parts, and the property of having a given volume is not something new that only emerges when we glue all the pieces together. In contrast, the topological properties of a given shape could be described as holistic properties, as it does not make sense to ask which part of a circle makes it a closed loop. Being closed is not a property of any one part of the shape; it is a property of the whole. Similarly, the Euler number of a shape is not something we can identify by looking at its separate, component parts: the way that the whole thing is glued together is absolutely fundamental. Indeed, if we do chop our shape into separate parts, the Euler number changes!

Chapter 13: LIVED EXPERIENCE AND THE NATURE OF FACTS

> 'Those who judge by employing a rule are in regard to others as those who have a watch are in regard to others. One says, "Two hours ago"; the other says, "It is only three-quarters of an hour." I look at my watch, and say to the one, "You are weary," and to the other, "Time gallops with you, for it has been an hour and a half." I laugh at those who tell me that time goes slowly with me and that I judge by imagination. They do not know that I am judging by my watch.'
>
> Blaise Pascal, 1623–1662

Rules and Reality

The word mathematics is derived from the Greek for 'teachable knowledge', and I think it is highly significant that when we are engaged with mathematical science, there is a distinctively close connection between what we know about a given subject matter, and what we can say about it. An insightful object of comparison can be found in Wittgenstein's *Philosophical Investigations*, where we are invited to consider the difference between 'knowing the height of Mount Everest', and 'knowing what a cello sounds like'. You cannot be credited with knowing the height of Everest without being able to tell someone. On the other hand, even a master cellist may flounder when asked to show that they 'know' what their instrument sounds like.

There are lessons from our contact with reality that cannot simply be packaged up and passed around. Not all human understanding is factual in form! On the other hand, the pursuit of scientific understanding is an essentially public enterprise, as a scientist's work largely consists in articulating the knowledge that is gained from investigating various objects. In the words of the physicist Niels Bohr, 'It is wrong to think that the task of physics is to find out how Nature *is*. Physics concerns what we can *say* about Nature.' In my view, it is the necessity of explicitly formulating and sharing our understanding that drives empirical science to the point where it becomes mathematical. To put it another way, nature could be there without us, but until we have constructed a language, there can be no facts. After all, we cannot have facts without a language to express them!

The language of mathematics is an integral part of the human adventure, and the history of maths tells us something about the cultures in which it developed. On the other hand, compared to other cultural artefacts, the world of mathematics is strangely timeless: you don't need to know how the Ancient Greeks lived to understand Euclidean geometry. The facts of mathematics can readily be transmitted from one culture to another, and as they have deep roots in our cognitive abilities, they seem to belong in every world. Indeed, in and of themselves, the facts of mathematics cannot tell us whether we are living in a world with one history or another. Whatever the past or present might be like, once we come alive to the possibility of using a form of mathematical language, we cannot imagine a world in which it would be impossible to think in those terms! For example, I might be utterly deluded about the world around me, but once I have learned how to count, I cannot imagine being part of a reality in which I cannot count.

As the quotation at the beginning of this chapter indicates, it is important to appreciate that mathematicians, scientists and engineers pursue their aims by employing rules. As a principal example of a rule-governed activity, consider the practice of counting. If a child does not count like the other children, we are right to tell them they are wrong, simply because they do not count like us. No other justification is required: we do not need to refer to a fact beyond the practice of counting itself. You might want to protest that there is more to the integers than the proper way to recite the counting song. That is no doubt true, but we should be careful, for as Wittgenstein observed, 'People can't distinguish the importance, the consequences, the application of a fact from the fact itself; they can't distinguish the description of a thing from the description of its importance.'

In short, mathematics is not a natural phenomenon, but it is experienced as though it were. For example, if I learn to count and see that $2+2=4$, I might want to insist that this isn't just true for me, or the people who count like me, it is a truth of the universe itself. Such a claim is eminently reasonable. Two stones and two stones do indeed make four stones, and wishing that it were otherwise cannot make a difference. However, by making this observation, I am not denying that mathematics is something we create. After all, it is us that see the stones as a group of objects, and a statement of fact is not the same thing as a physical state of affairs. To put it another way, our versions of how the world works do not write themselves, even if our accounts are genuinely faithful to the reality in which we find ourselves.

In my view, mathematical truth comes into being with the rule-governed use of symbols: without the explicit rules or principles set down by ourselves or our predecessors, there can be no mathematical facts. On the other hand, mathematics is more than merely shuffling symbols according to a given set of rules, because once we have a rule-governed system that we can gainfully employ, we can view the world though the logical lens provided by that language. And once a person uses a word or concept, it acquires real meaning. This image of people using symbolic systems might suggest a picture of mathematical factuality that is comprised of two parts: the actual, meaningful, historical concepts and symbols used by physically embodied human beings, together with the ahistorical or timeless symbolic facts which could in principle be verified by any suitably programmed computer.

This two-fold image of mathematical truth has a lot to recommend it, provided that we acknowledge the impossibility of seeing a dividing line between the meaningful mathematics that we know and use, and the computable, symbolic systems that are consistent and legitimate, but yet to be discovered. After all, the moment a person talks about a possible computable system, it becomes a system that has actually been used! As Wittgenstein remarked, 'There is a feeling: "There can't be actuality and possibility in mathematics. Everything is on one level. And in fact, is in a certain sense actual." And that's correct. For mathematics is a calculus; and a calculus does not say of any sign that it is merely possible; rather, a calculus is concerned only with the signs with which it actually operates.'

Calculations and mathematical arguments can be used to understand the world, and that is what makes mathematics profoundly meaningful. On the other hand, mathematical facts are not contingent on physical states of affairs, as the language of maths deals in generalities, and is not supposed to be understood by reference to specific or historical events. In contrast, scientists make claims about the empirical world, and completely different theories can all refer to the same empirical phenomena. Consequently, scientists have to accept that new evidence or a new approach might show that their theories were wrong. Mathematicians often have hunches that turn out to be wrong, but the results that they prove acquire a certainty that science will always lack, precisely because mathematical truth is always truth within a given system. We might say that our language enables us to look at the world through a particular mathematical lens, and however our surroundings may change, we are free to keep our logical glasses just the same.

Perhaps the most important thing is that mathematicians articulate their concepts in a way that enables deduction. This fundamental discipline constrains and shapes the body of mathematical facts, but there is more to mathematics than a tower of deductions. In short, mathematicians investigate describable, conceptual schemes. The same might be said of other theorists, but the logical structures of mathematics are distinctive because the point of interest is nothing beyond our own definitive statements and symbolic forms of practice (e.g. counting). Mathematicians are free to count actual objects, but unlike other forms of theory, the terms of mathematics don't need to refer to any external objects. I therefore conclude that the objective reality of mathematics is not activity in the brain, or some magic realm of mathematical forms. Like Pascal's watch at the beginning of this chapter, what matters are the *statements* that a mathematician *employs*.

Although few people focus their mind on the purely mathematical, I believe that our natural capacity to understand such truth is a vital part of what it is to be human. This is true of everyone, not just the well-trained experts. For example, although children need to be initiated into using the vocabulary of straight lines, squares, triangles and other specific shapes, our capacity to recognize the sense of such a language is innate, and we cannot imagine looking at the world without finding shapes within it. This is not to say that human nature is explained by a hidden mathematics, but we are deeply moved by the truly obvious, and our mathematical traditions begin with statements of the obvious. I would also say that when a person makes the effort to assimilate the corpus of maths, they encounter the fact that as rational beings, we are capable of experiencing truth. This is a fundamental fact of human nature, and it is inseparable from our use of language.

To account for the vitality or meaningfulness of the symbols that they use, many mathematicians and some philosophers are Platonists. Platonists believe that mathematicians study abstract objects that exist independently from our means of studying them. I find this claim of independent existence unconvincing and ultimately hollow, because it seeks to legitimize the things that people actually know how to do (e.g. doing sums or writing certain kinds of proof) by appealing to something that is beyond our knowing (that is, an imagined domain of transcendent abstract objects).

As we'll see in the following section, the status of mathematical objects is a delicate matter, but in practice it seems that it doesn't really matter what mathematics is 'about'. Whether we imagine some pieces of string on the muddy fields of Egypt or some perfect timeless shapes, the practice of deducing the areas of shapes ultimately hinges on the language we employ, not the pictures we have in our minds. We might imagine our subject matter in many different ways, but once we describe the object that we wish to measure, it is the stated lengths and angles that matter, and the possibility, or impossibility, of using such descriptions together. More generally, what matters in mathematics is how we can work from one statement to another, not the objects that we think we might be talking about.

Mathematicians are free to study anything that can be

identified and characterized by means of a consistent description. Chess, sudoku, and other rule-governed games are essentially mathematical, though a game like chess is much more arbitrary than the structures that mathematicians choose to study. I say this because the concepts that mathematicians investigate are closely related to a large complex of other concepts, including ideas that are used in science, or everyday accounts of what the world is like. The interconnectedness of mathematics is truly profound, and the subtle relationships between ideas provide a way of valuing some mathematical insights over others. By way of analogy, if a historian studies a particular event, it is because they think that the event in question has bearing on events in general, and an equivalent comment holds true for mathematicians.

If we are interested in solving a particular problem, or understanding how different mathematical ideas are related, it is certainly not the case that studying one set of consistent axioms is just as good as studying another. It is not like choosing between chess and checkers! Using given axioms to generate new theorems is indisputably important, but mathematical progress also involves the formation of new hypotheses and new conceptual frameworks, and novel forms of mathematics can be judged by the light it sheds on previously solved and unsolved problems. My point is that even if every formal approach is equally mathematical, some arguments are more memorable than others, and not all of mathematics is deemed to be worth remembering.

The Objectivity of Maths

What is wrong with saying that mathematicians study abstract objects? After all, hundreds of generations of mathematicians have studied patterns in the integers. Surely it is safe to say that the integers are abstract objects? People imagine the objects of mathematics: is our imagination the realm of maths? But don't the integers possess their own form, regardless of what we think? Would they be there without us? Is there a timeless, spaceless portion of reality where mathematical objects exist?

If we are not careful, these bewitching images can stop us from looking at the things that we actually know. We do not need access to timeless, perfect objects. Instead of casting meaning away from humanity to some ultimate, all-seeing judge, we can suppose that mathematical language brings about the reality it declares. Rather than thinking of mathematical objects as the underlying bedrock of maths, I think we should concede that the irreducible fact of the matter is that people are capable of employing the language of mathematics. In particular, the brute fact about numeracy is that in the medley of a lifetime we have assimilated the basic concepts of the integers, and possess the ability to employ them. As a result, people really can count, and are capable of making true, arithmetic statements. So, you might ask, if we experience truth by 'assimilating mathematical language', are the facts of mathematics invented or discovered?

The advent of novel mathematics is a unique kind of event, but I am inclined to say that 'invent' is the better metaphor. Like engineers building a conceptual device, mathematicians need to be inventive, but we discover that our invention works, and is intelligible to our colleagues. By way of analogy, we invent the rules of chess, but the game has a certain autonomy, and given its rules we cannot change the fact that forcing checkmate with two knights is impossible. Indeed, we might well say that we discover this fact about our own invention. Likewise, mathematics is a human creation, but we cannot simply bend our creations to our will, as mathematical objects and the relationships between them are constrained by stated principles. As in the case of chess, the partial autonomy of mathematics derives from the fact that our creation is governed by rules that we can state.

In short, mathematics is a language, and languages are cultural artefacts. This is a subtle claim, as it does not mean that the objects of mathematics are arbitrary inventions, or only exist 'in our head'. We can find meaning by making a language our own (a process that enables us to form certain kinds of thought), but language itself does not belong to any individual, or exist in any one brain. By its very nature language is sharable, with deep roots in observable patterns of behaviour, so although we need brains to think, we do not understand what sentences mean by looking at each other's brains! We may feel as though our thoughts are private, and contained within our own minds, but mathematical structures and symbols are essentially public and sharable, even if our feelings about them are not. As Heraclitus of Ephesus remarked twenty-six centuries ago, 'Although the forms of reason (*logos*) are shared, most men live as though their thinking were a private possession.'

We encounter the objects of mathematics through our use of language (particularly the terms used in measurement), and it is the use of rule-governed language that can bring such objects to mind. Furthermore, our understanding of number is very deep rooted indeed, because we naturally appreciate that objects can be added to a group, or placed into containers. Indeed, when we are teaching arithmetic we quite reasonably take it for granted that children will be able to understand the process of adding or taking away objects from a collection, even if they don't yet understand the symbols of arithmetic. By building on the highly intuitive, even pre-linguistic concept of a collection, we have been able to form highly robust abstract number concepts. Such abstract concepts require the use of rule-governed language, and as such they have an essentially cultural component.

For example, we are justified in saying that there 'is' one and only one number '3' precisely because it is actually possible to count to three, and a public can assess whether or not someone has done this correctly. The word is *meaningful* because we have an intuitive understand of collections of countable objects, but the facts about numbers cannot be separated from the conventions for their use. To put it another way, we should only accept a set of axioms if they are in accord with the ideas that we associate with our symbols, but it is the fixed axioms and other rules governed principles that make mathematics precise, stable and sharable. Furthermore, once we accept a set of rules and begin to employ them, we can discover the patterns that are drawn out by those rules.

Because mathematics is precise, stable, sharable and rooted in human cognitive abilities, the objects of mathematics have many things in common with nameable, physical objects. However, numbers and other mathematical objects are unlike physical objects in that the existence we can claim for the integers (say) is not an independent, isolatable one. In other words, number words resemble names, but it is misleading to take the object '3' too literally. We do not need to imagine some transcendent object that has all the properties of three-ness, and none of the mess or irrelevance of a particular system of notation. All we need is a series of number words, so we can start to count. That is to say, the truly essential point is that each number stands in a structured relationship to the other numbers. Not only can we identify an item in a particular sequence as 'being the third', it is also the case that each integer can be characterized and identified by its place within a definitive list of the integers. In that sense number words are very much like names. However, the essence of a 'place' is nothing other than its relation to the other places.

For example, there is no special quality to being first in line other than the definitive fact that being first means being before the other places, without which there could be no 'first'. As Stewart Shapiro wrote in *Thinking about Mathematics*, 'Any small, movable object can play the role of (i.e. can be) black queen's bishop. Similarly, anything at all can "be" 3 – anything can occupy that place in a system exemplifying the natural number structure.' It would be absurd to try and make sense of the term 'black queen's bishop' without referring to the chess board, and likewise, we cannot make sense of any particular number without reference to our number system. In short, the essence of an individual number is not something intrinsic to an individual abstract object (whatever that might be): the properties of numbers make sense only in the context of a number system. For example, we know that '3 is odd', but that does not mean that oddness is a property intrinsic to some special object known as 3. Another way of saying '3 is odd' is to say that 'there is SOME integer n such that $3 = 2n + 1$', so the oddness of the number 3 precisely consists of a relationship between the number 3 and some other numbers.

In short, the specific rules, definitions, and notational systems that we adopt are central to mathematical practice, as it is these things that specify the game that we are playing. However, that does not mean that mathematics is 'a game played with meaningless symbols', as the formalists suggest. Our choice of rules is motivated by underlying concepts, and although the things that computers can do are central parts of mathematics, there is much more to the subject than the valid application of formal languages like PC. When I say that the irreducible fact of the matter is that mathematical arguments exist, I am not only thinking of computer checkable proofs and calculations. I am also thinking about the informal arguments, images and metaphorical statements that dominate this book, and the broader play between various kinds of proof and demonstration.

Euclidean geometry is often described as a great inverted pyramid: a vast tower of deductions resting on a handful of explicitly stated principles. Despite the truth in that seductive image, the development of mathematics as a whole is much more than the orderly application of deductive principles, with one statement following from another. For starters, creative mathematicians not only work forwards from axioms to theorems. They also work backwards, starting with a problem and formulating hypotheses that might enable them to deduce solutions to the problem at hand. Indeed, as is the case with Euclidean geometry, we often realize that our foundations are good foundations precisely because a given conceptual scheme enables us to tackle effectively a given set of problems.

My point is that creative mathematicians don't just make obviously legitimate deductions: they try to spot patterns, they make conjectures, they break problems into smaller sub-problems, and they guess at general statements after examining a few specific instances. They also try to solve problems by employing analogies, as progress is often made when someone says, 'Maybe we can understand this problem by thinking in the terms we use to solve that other problem.' These features of analytic thinking are common to many fields of endeavour, and with experience a skilled mathematician learns to put their finger on the critical point.

Mathematicians make progress by thinking about the challenges that face the mathematical community, and this involves intuitive leaps of understanding, as well as formal, deductive practice. It is certainly possible to develop mathematical intuition (you just need to spend some time doing mathematics), but mathematicians wisely trust their rules and symbols, not their imaginations. Working mathematicians may harness all manner of mental faculties, but the distinctive discipline of mathematics is to aim at making the relevant deductions completely formal and explicit, so that others can follow the argument.

Meaning and Purpose

Nobody seriously believes that one piece of mathematics is as good as any other, even though a dull or arbitrary piece of maths is every bit as correct as the most celebrated proof. Mathematicians are motivated to engage in the research that they do, and they have reasons for employing the rules that they decide to follow. Although our feelings towards mathematical statements cannot change the facts themselves, our motivation or sense of purpose in doing mathematics is absolutely essential. After all, people have reasons for doing what they do, and that is an essential part of the mathematical experience.

People are motivated to engage with mathematics for all kinds of reasons. Important practical concerns require mathematics, people want to solve famous problems, we might wish to show connections between historically separate theorems. Most importantly of all, mathematical patterns can simply capture our imagination, provoking us to thought. As the mathematician G. H. Hardy perceptively remarked:

> There are many highly respectable motives which may lead men to prosecute research, but three which are much more important than the rest. The first (without which the rest must come to nothing) is intellectual curiosity, desire to know the truth. Then, professional pride, anxiety to be satisfied with one's performance, the shame that overcomes any self-respecting craftsman when his work is unworthy of his talent. Finally, ambition, desire for reputation, and the position, even the power of money, which it brings.

Formal definitions and computer-checkable deductions are absolutely central to mathematical science, and it is obvious that computers do not need to have a sense of the motivations of mathematicians in order to do what they do. Nevertheless, motivations are an essential part of the mathematical experience, because the satisfaction of a compelling argument does not simply consist of its publicly established validity. My point is that the vitality of computer-checkable, symbol-based methodologies is not inherent to the rules alone, but rather depends on the endlessly mysterious and unruly process whereby any kind of symbolic representation may come to engage with our imagination. To put it another way, we may have no choice but to *define* a word in its relation to other words, but our sense of *meaning* is essentially bound to the apprehension of human purpose.

It is well known that mathematicians rely on clearly stated definitions. Consequently, we might be inclined to think that motivations are unimportant in maths because our feelings about the facts cannot change the facts themselves. However, that kind of insensitivity to underlying motivations is far from being a unique or distinguishing feature of mathematical science! It is physically inevitable that people do not need to share the same sense of purpose in order to use the same language. Indeed, even our language for speaking about our own feelings or emotions cannot be any better than our language for speaking about other people's feelings, despite the fundamental asymmetry between our knowledge of self and other.

In the particular case of mathematics, the facts themselves can be presented or represented by a formal scheme, and we don't need to refer to people or their cultures. But there is more to the practice of mathematics than formal definitions! As well as the statements of mathematics, there are also mathematicians, who are cultured people that find a palpable weight in their use of symbols. As is the case with all kinds of language, we can sketch out definitions in the abstract, but meaning can arise only in the context of life, when flesh and blood human beings engage the world with words.

People use mathematical concepts, and in my view our very sense of reality is shaped by the fact that our words can satisfy (or fail to satisfy) the motives that we hold. It is manifestly the case that linguistic expressions can be effective in ways that we ourselves are in a position to confirm: we have enough in common for language to work. For example, we might want to be successful in a highly instinctive game called 'name that object', and more sophisticated motives can likewise be fulfilled.

When it comes to assessing another person's mathematical work, we look to our definitions to answer the basic question: 'Has the ritual been performed correctly?' However, mathematics involves much more than simply using symbols in the same way as our peers, as we should also consider the radically mathematical question, 'What is essential to the workings of my ritual, and what is simply arbitrary?' Fixed rules unfold as they must, but maths is more than a correctness of symbolic form, as such languages may be fit for human purpose. In particular, the invention of number words has given us a linguistic technology that is genuinely capable of elucidating plurality: a presumably ancient goal.

Our sense of mathematical purpose can be truly deep and beautiful: grounded in the obvious, but endlessly otherworldly. As I hope my book has shown, mathematical arguments can be deeply striking, and they play a crucial role in our best attempts at comprehending the world. Furthermore, as esoteric as it may seem, I am convinced that the philosophical contemplation of mathematical practice is profoundly worthwhile. A shift in our philosophy will not change the facts themselves, but our attitude towards the factual helps to shape our lives, and what could be more important than that? With that point in mind, I cannot think of a saner final comment than these lines from *Circles*, by Ralph Waldo Emmerson: 'Every ultimate fact is only the first of a new series. ... No facts are to me sacred; none are profane; I simply experiment, an endless seeker with no past at my back.'

Further Reading

The following lists are arranged by subject matter, with more accessible books placed at the beginning of each list, while more scholarly tomes are at the end of each list.

Introductions to Higher Mathematics

Alex's Adventures in Numberland, by Alex Bellos (Bloomsbury, 2010).

Imagining the Numbers (Particularly the Square Root of Minus Fifteen), by Barry Mazar (Penguin, 2003).

Mathematics: A Very Short Introduction, by Timothy Gowers (Oxford University Press, 2002).

The Mathematical Experience, by Philip J. Davis and Reuben Hersh (The Harvester Press, 1981).

What is Mathematics? An Elementary Approach to Ideas and Methods, by Richard Courant, Herbert Robbins and Ian Stewart (Oxford University Press, 1996).

Concepts of Modern Mathematics, by Ian Stewart (Dover Publications, 1995).

From Calculus to Chaos: An Introduction to Dynamics, by David Acheson (Oxford University Press, 1997).

Histories of Mathematics

Number – The Language of Science, by Tobias Dantzig

(Macmillan Company, 1930, reprinted by Pearson Education, 2005).

A History of Mathematics, by Carl Boyer (Wiley, 1991).

Euclid's Window, by Leonard Mlodinow (Penguin, 2001).

Mathematics and Its History, by John Stillwell (Springer-Verlag, 1989).

Pi in the Sky: Counting, Thinking and Being, by John Barrow (Penguin, 1993).

Greek Mathematical Thought and the Origin of Algebra, by Jacob Klein (Cambridge University Press, 1968).

The Crest of the Peacock: Non-European Roots of Mathematics, by George Gheverghese Joseph (Penguin, 1991).

A Concise History of Mathematics, by Dirk Jan Struik (Dover Publications, 1987).

Mathematical Thought from Ancient to Modern Times, by Morris Kline (Oxford University Press, 1972).

Philosophy of Mathematics

Thinking about Mathematics, by Stewart Shapiro (Oxford University Press, 2000).

What is Mathematical Logic?, by J. N. Crossley, et al. (Dover Publications, 1972).

Naturalism in Mathematics, by Penelope Maddy (Oxford University Press, 1997).

Proofs and Refutations, by Imre Lakotos (Cambridge University Press, 1976).

Where Mathematics Comes From, by George Lakoff and Rafael Nunez (Basic Books, 2000).

Philosophy of Mathematics (Selected Readings), edited by Paul Benacerraf and Hilary Putnam (Cambridge University Press, 1983).

18 Unconventional Essays on the Nature of Mathematics, edited by Reuben Hersh (Springer, 2006).

Reasons Nearest Kin, by Michael Potter (Oxford University Press, 2000).

Tractatus Logico-Philosophicus, by Ludwig Wittgenstein. I recommend reading this with *Wittgenstein's Tractatus*, by Alfred Nordmann (Cambridge University Press, 2005).

Philosophical Investigations, by Ludwig Wittgenstein (Wiley-Blackwell, 4th Edition, 2009). I recommend reading this with *Wittgenstein's Philosophical Investigations: An Introduction*, by David G. Stern (Cambridge University Press, 2004).

Wittgenstein's Lectures on the Foundations of Mathematics (University of Chicago Press, 1939), *Remarks on the Foundations of Mathematics* (Blackwell, 1978), and *On Certainty* (Blackwell, 1993) are also well worth reading.

Symmetry, Geometry and Topology

Finding Moonshine, by Marcus du Sautoy (Harper Perennial, 2008).

Fearful Symmetry, by Ian Stewart and Martin Golubitsky (Penguin, 1992).

The Ambidextrous Universe, by Martin Gardner (Penguin Books, 1964).

Mathematics and Optimal Form, by Stefan Hildebrandt and Anthony Tromba (Scientific American Books, 1985).

Symmetry, by Hermann Weyl (Princeton University Press, 1952).

Gödel's Theorems and Related Results

Gödel, Escher, Bach: An Eternal Golden Braid, by Douglas R. Hofstadter (Basic Books, 1979).

Gödel's Proof, by Ernest Nagel and James R. Newman (New York University Press, Revised 2002).

Forever Undecided, A Puzzle Guide to Gödel, by Raymond Smullyan (Random House, 1987).

On Gödel, by Jaakko Hintikka (Wadsworth Publishing, 1999).

Mathematical Biology and Emergent Phenomena

Nature's Numbers, by Ian Stewart (Phoenix, 1998).

Emergence, by Steven Johnson (Penguin, 2001).

The Self-Made Tapestry: Pattern Formation in Nature, by Philip Ball (Oxford University Press, 2001).

On Growth and Form, by D'Arcy Thompson (Cambridge University Press, 1961).

Models of Biological Pattern Formation, by Hans Meinhardt (Academic Press, 1982).

Acknowledgements

My way of thinking about thinking has been through mathematics, but I have also been influenced by my father's love of Wittgenstein. I owe countless intellectual debts, but I owe particular thanks to my friends and family for their love and support. I would also like to thank John Woodruff and Hugh Barker for their helpful comments, as well as the many mathematicians, historians, philosophers and popular science writers who made this book possible. For the sake of clarity and brevity I have not always been able to credit the people whose work has influenced my own, and I would like to apologize to anyone who feels slighted by such an omission.

I might be described as a philosopher, a mathematician or a theoretical biologist, but I would not claim to be a historian. There is a list of mathematical histories at the back of this book, and I would also like to recommend an excellent online source: the University of St Andrew's website – www-history.mcs.st-and.ac.uk. Any historical errors are my own, but if my account is faithful to actual events, it is thanks to the scholarship of others. Finally, I would like to dedicate this book to the memory of James Wilson: he helped me stop and question what was obvious.

Index